KU-542-384

PELICAN BOOKS

A326

MICROBES AND US

HUGH NICOL

MICROBES AND US

HUGH NICOL

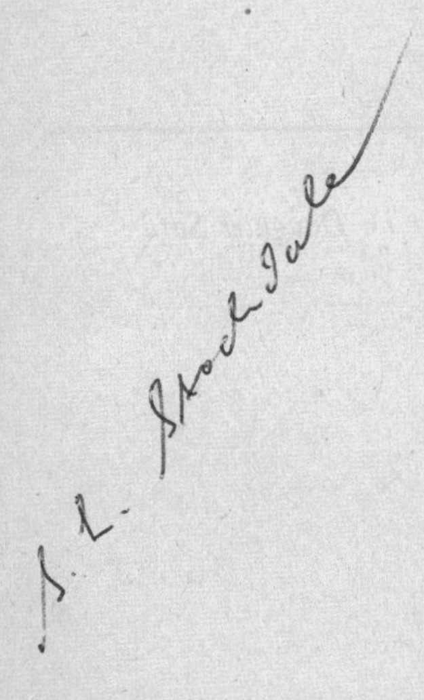

PENGUIN BOOKS

Penguin Books Ltd, Harmondsworth, Middlesex

U.S.A.: Penguin Books Inc., 3300 Clipper Mill Road, Baltimore 11, Md
[*Educational Representative:*
D. C. Heath & Co, 285 Columbus Avenue, Boston 16, Mass.]

CANADA: Penguin Books (Canada) Ltd, 47 Green Street,
Saint Lambert, Montreal, P.Q.

AUSTRALIA: Penguin Books Pty Ltd, 762 Whitehorse Road,
Mitcham, Victoria

SOUTH AFRICA: Penguin Books (S.A.) Pty Ltd, Gibraltar House,
Regents Road, Sea Point, Cape Town

—

First Published 1955

To Daughter Sara

Made and printed in Great Britain
by Western Printing Services Ltd
Bristol

CONTENTS

PROLOGUE

Mr. Pott receives four species of earths;
A Manual of Chemistry, translated from the French of M. Beaumé,
by John Aikin, Surgeon: Warrington, 1778

About Mr Pott I admit to knowing nothing whatever. He must have been a mid-eighteenth-century philosopher, but I am ignorant of even his nationality. I have not troubled to find out anything more about him than what Aikin translated about his views on the nature of earths.

This book is principally about air, and about the instrumentality of living things, and of microbes in particular, in using air, and combining it into solid form in such ways that, so far and speaking approximately, there has been a balance between food-supply and human population. That theme is enough for any one book about earth in the sense of the soil and its microbes. What space there is to write about earths in the eighteenth-century sense I allocate now to a warning about the very great importance of earths in present and future utilization of air by man in co-operation with the microbes of soil.

Mr Pott's earths were the calcareous, the gypseous, the argillaceous, and the vitrifiable. He did not receive the combustible. Beyond reasonable doubt, he rejected that as earth, because he regarded an earth as something which could not be burnt. On that point I am broadly of the opinion of Mr Pott. Yet it is important to indicate here that man's technical conquest of the earth (about which he is so proud), and his quest for food, have, since Mr Pott's time, come to be coterminous with man's ability to burn the combustible fractions of the planet.

Living upon sources of energy which were renewable each year, and struggling and dying in accordance with the limitations thus set, Mr Pott's contemporaries were perhaps not in a better situation than we are; but they were in a profoundly different state of equilibrium with their world. It was the same kind of balance as had existed since man began; the only difference from what had gone before was – as we can see now – that it was approaching its end, and was about to give way to an era (ours) which called up its own technical uncertainties.

One thing is certain: the importance of Mr Pott's four earths has not changed in respect of food-production since his time.

In this book I am not trying to write a treatise on fertilizers or soil amendments, or to examine, unless by inference, the pressure of population. Every literate person can do simple arithmetic. If population increases by one per cent per annum, a sufficiently good working answer to the implied problem of population can be obtained for the next twenty years, whether the compound-interest rule is used or not.

However, the principles of soil microbiology are less well known, and not less inexorable. That is sufficient justification for this book.

Many people point to mineral resources as reasons for not worrying overmuch about the next fifty years. In England the deposits of gypsum are vast – perhaps greater than the reserves of coal. Given a sulphur shortage, there might be no harm, from an insular point of view about food production or the manufacture of rubber goods, in converting all the British gypsum into sulphur: at least up to the limit of the coal available for that chemical purpose of transformation of earths. To do so is possible nowadays.

There is looser talk about the potentialities of irrigation overseas, and the prospects of reclaiming deserts and saline lands.

Suppose an area of fifty thousand square miles of saline land to be marked out as being flat and otherwise cultivable if reclaimed. It is part of the simple common sense of soil equilibrations that saline lands tend to occur in arid regions where the net amount of water is insufficient. If the soils were moist under what we in Britain regard as ordinary amounts of rain, the salts would not have accumulated, or would have been washed out. For countering excess salts, whether in arid soils or in those flooded by sea water, two of Mr Pott's earths are specific, and the chief of these is the gypseous.

In our saline area there will probably be no water at hand. If water is to be produced by distillation or other secondary means, that would necessitate a colossal consumption of energy. Suppose the water problem overcome; then suppose a single dressing of ten tons of gypsum per acre to be satisfactory for redeeming the soil from salinity. The tonnage of gypsum would run into nine figures; as does the annual output of coal by Great Britain. Divide that tonnage of gypsum by three, and it would still represent a big hole in even the largest deposit. And what of the fuel needed for grinding and transporting the mass? If power for those mechanical ends is supplied by tomic energy (I follow Professor Sir Frederick Soddy in disliking the misnomer 'atomic' in such connexions) instead of by use of combustible earth, there will remain an inescapable need for fuel to make fertilizers.

How many millions of people could be fed from that much new land, after all that effort, and supposing the fuel demands to continue to be met?

About twenty.

That is an outline for discussion of some merely chemical aspects. As things are, the chemical aspects should surely be better understood, and more easy to ventilate, than are the microbiological implications and consequences of schemes for greater production of food. I think this book should serve a purpose. It may rouse the planners and technologists from believing that all is well if they dream in terms of world mineral resources, while making propaganda for greater use of fertilizers, wider adoption of higher-yielding crops, and the rest of the alleged escape routes. None of such dreamers and arithmeticians can know anything useful about microbiology and its relation to living.

People and animals cannot be fed by arithmetic; nor can it be guaranteed that people can be kept at peace out of an acquaintance with soil microbiology. The least, and perhaps the most, that I can say is that there can be no sense in supposing that we can go on living except by conforming to the cycle of life.

Therefore, the first thing to be done before assessing our prospects is to comprehend the cycle.

Production of food from the land – with which I am chiefly concerned – is limited by a number of things, not one of which can be said to be well known. Thus, while every educated person knows something about the way in which plants make use of the energy of sunlight, hardly anybody really comprehends the next step needed to put facts about photosynthesis into relation with production of food by farming. As one result of this lacuna, most people think that discovery of a suitable method for increasing our present low efficiency in capturing the energy of the sun would open a fairly easy way to support a larger population. Irrigation and artificial rain-making are favourite 'specifics' for dry areas; but how widely is it understood that all agriculture of humid regions depends on control of water and is limited by the extent to which water in all forms can be utilized and controlled?

In most discussions about food-production a good deal of misinformation is common property; there is not just a lack of information and inadequate linking-up of what is known. Everyone who knows anything about diet knows the value of proteins; but who except soil microbiologists and marine biologists know that the supply of proteins is decided by micro-organisms; and who except one of those small classes of people know that the amount of protein obtainable from the land is limited by one special kind of bacteria?

To understand the cycle of life well enough to *see* what it means for production of food, it is necessary to possess the master-facts about microbes, soil dynamics, botany and plant physiology, human and animal nutrition, and chemistry. These seem to be a pretty miscellaneous lot; and a reader might throw in his hand before attempting a synthesis of so many apparently assorted branches of knowledge. He might say that he knows something about some of them, and he may even be a specialist in one of them, or in medicine or one of the sciences related to them. But is all that he has learnt about dietetics or farming or botany or bacteriology, wrong?

By no means; but it has almost certainly been so far incomplete, or so strangely lop-sided, that it does not help to grasp the main principles together. Bacteriology (for example), as it has been taught and learnt, may have been related almost exclusively to disease, and not to living; or chemistry, as set out in the text-books, can have had some of its fundamental principles quite overlooked whenever a food topic came within range. Chapter Twelve gives one instance of this: another is to be found in the chemical principle of oxidation and reduction applied to food itself. We must oxidize food in order to live. It follows that some kind of combustible earth, such as coal, must be burnt if we are to have more food, or if there is to be food for more of us. What chemist thinking about food-production will recall, in this connexion, that the route by which a substance is made is chemically unimportant? But that being so, it is essential to use up some of our limited stock of fossil materials if we want to have more carbon and nitrogen combined into foodstuffs. Consequently, whether food is made synthetically from coal, or is prepared by less 'futuristic' methods based on farming, the demand on combustibles will be about the same; though, of course, the routes may differ in efficiency, as the products may differ in palatability.

The science which includes the study of principles of food-production is called by the curious name 'agricultural chemistry'. That is a misnomer; and I do not wish to go into the history of the name, but to present a few of the consequences of relegating the understanding of conditions of our existence to a subject with such a bucolic name.

One consequence is that many people look upon 'agricultural chemistry' as a poor relative of real chemistry. This view is strengthened when it is recalled that my subject leads to the production of no gadget and nothing like a dye or a drug which can be confined in a bottle and exhibited as a triumph. The technologies of fertilizers and of vitamins, for example, lie outside 'agricultural chemistry' – except that agricultural chemists have to know enough about such subjects to give a smattering of them to elementary students. The circumstance that hardly anybody except students of agriculture comes within hail of a course of 'agricultural chemistry' is another reason why comprehension of socially important principles of chemistry and microbiology is not widely diffused. How little the subject is esteemed may be gathered from the abolition of a Chair of Agricultural Chemistry in England during 1953.

What is called 'agricultural chemistry' has nothing necessarily to do with agriculture or food. It is a science of ideas linking understanding of the workings of soils and plants with the needs of man.

Lecture-courses lasting several hours could be given on some topics of 'agricultural chemistry' – for instance, preservation of food, soil fertility, or the principles of life itself – without mentioning any chemical formula more difficult than H_2SO_4 (which I presume I need not explain here). This book, in effect, does that for several such topics. Thus, no reader willing to acquire new ideas and cast out old ones need be deterred through expecting that a vast acquaintance with science is required of him.

So few are agricultural chemists and their opportunities for presenting ideas to the public at large that most of the linkages of, and deductions from, the simple facts given will be fresh to most people; including the 'real' chemists and the declared social scientists. Thus, the lay reader need not be humble about the supposed imperfections of his mental scientific equipment: on this occasion he will be tackling a scientific subject by starting at the same level as almost everybody else.

Once the principles are understood so as to gain chemical proportion in a microbiological matter, it immediately becomes possible to see why we are all in a perilously artificial relationship with the soil microbes.

CHAPTER ONE

A Saltspoonful of Soil

'I should see the garden far better,' said Alice to herself, 'if I could get to the top of that hill: and here's a path that leads straight to it – at least, no, it doesn't do that – ' (...) 'but I suppose it will at last. But how curiously it twists!' – *Through the Looking-Glass*

OURS is a busy, vexed, quarrelsome world. It contains about two thousand million people intent on gaining a living, on learning, on loving, on killing, on inventing, on changing, and on drifting. Every one is occupied with his or her ego. Conflicts arise, and stories of the more sensational human conflicts are received in the newspaper offices, and are called news. The journalist pacing the pavement of Fleet Street knows that behind the façades of the big newspaper offices are telephones and teleprinters receiving news from the whole world.

There is no soil in Fleet Street, only pavement. If you were to pace the fields or the moors instead of a dead pavement, your every footstep would cover a population many times greater than a couple of thousands of millions. There is at least that number of living things in a saltspoonful of soil. They teem, eat each other, and have their ups and downs in a perpetual conflict for food and existence.

These creatures range in size from beings just visible to the naked eye, down to those which are about 1/20,000th of an inch across and can only be seen with a powerful microscope. But, though small, they are alive; they are involved in the struggle for existence – in our struggle as well as theirs – and they exist in a perpetual conflict. Those reasons both make their study interesting. Change – a new balance of power – a fresh dominance – news, in fact – is always being presented. But not by teleprinter; microbe news has to be recorded in other ways.

The method used for study of the modes of life of microbes belongs to the sciences collectively called microbiology. The name is perhaps not so very good; it suggests a small biology, whereas it really means the biology – the life-study, that is – of small things.

If at this stage you were to ask me to say what a microbe is, I could give no better definition to the term than by saying that strictly speaking a microbe is any small form of life. For some purposes of discussion of food-production it is convenient to include among the microbes of soil the various multicellular animals up to and including the earthworms or even larger forms. Microbe in the sense of micro-organism, of which there can be several millions in a salt-spoonful of soil, is a form of life so small that a microscope has to be used to make it visible.

Even so, some exceptions have to be apprehended. The egg of a mammal, such as a human being or a cat, is so small that it cannot properly be seen by what the German language calls the unweaponed eye; the male equivalent of such eggs is still smaller, and is truly microscopic. Small, also, are pollen grains, which are the male 'eggs' of plants. Such structures as these are not considered to be microbes, and possibly you will understand the reason without my stressing the point about an independent existence. Microbes (micro-organisms) are very small Peter Pans: they never grow as individuals beyond a minute size.

When microbiologists talk about growth of microbes, they almost always mean an increase in numbers. In general, therefore, microbial growth is to be understood as multiplication, leading to an increase in population of one or several species.

In all respects, microbes are neutral. Like Alice, they never ask advice about growing. This is not because they are too proud, but because the growth in numbers of the individuals of any one species, or the balances between numbers – and consequently the relative importances – of different species is determined wholly by the conditions to which the species is or are exposed.

In soil the conditions can vary extensively from moment to moment and from place to place. The microbial population of soil adjusts itself exactly, as regards numbers of individuals and balance of effectively operating species, to whatever the prevailing conditions may be. All species of micro-organisms are actually or potentially present in any bit of soil. They survive and are active in proportion to the extent to which conditions suit them. An appreciation of that fact will lead to a grasp of the other and useful fact that what determines the activity and continuance of operational strength of microbes is the condition of the environment, and not – unless in special circumstances – the mere presence of a species or its accidental or intentional introduction by man.

This ability or failure to grow, depending on whether the *conditions* are suitable or not, is not peculiar to micro-organisms. It is also exhibited by weeds and cultivated plants, for instance. Sowing the seed will not ensure the appearance or survival of a plant if it is subjected to competition from other kinds under conditions not favourable to the newcomer. If the introduction is to flourish, the right conditions for its growth must all be provided. The presence of a seed, or mechanical implantation in the soil, is only one of those conditions.

This business of living in competitive balance with other kinds of organisms is the normal and natural mode. In agriculture we make some kinds of plants and animals depart very considerably from what is natural, and we succeed only to the extent that we artificially provide a new set of conditions temporarily favourable to new associations of plants – and of soil microbes.

I believe that a study of soil microbes – not as single species to have their characters described and illustrated as if they existed in isolation, but as an essay in equilibria – is the most highly philosophical branch of microbiology. Indeed, it is not a branch, but the system of roots. Other branches of microbiology, such as those concerned with the specialities of diseases associated with presence of micro-organisms, or with the culture of yeast or the making of penicillin, are but twigs or fruit on the microbiological tree: the tail, rather than the legs, of the corpus of biological endeavour. Their study may be said to be of the nature of a craft rather than a philosophy.

It is often thought that it is necessary to add organic matter to soil in order to feed the soil microbes. This comes from a nearly total misunderstanding. The organic matter added to soil and subjected thereby to microbial attack has many valuable qualities, of which the physical are probably the chief and the plant-nutritional the next important. These qualities arise largely from the actions of soil microbes before and after the dead matter is added to soil. Nevertheless, there is no point whatever in feeding soil microbes in the expectation that they will profit from feeding in the way a pet cow or any other individual organism benefits individually and to the delight and profit of its rearer. That idea can come only out of notions derived from contemplation of individuals reared in isolation, or maybe in herds, but in any case severely protected from the exercise of natural equilibrium. The idea shows ignorance of the concept of balance and equilibrium in accordance with conditions – of which food-supply is one.

The microbes are neutral. Given organic matter or other food, in any quantity and of any kind and in accordance with any set of other conditions, the soil microbes will immediately act to adjust themselves to the new disturbance of equilibrium represented by the accession of food. In imagining their adjustment, it is helpful to think, not of the totality of 'the microbes', but of one or more kinds specially able, for a short time, to utilize the new food. It will be remembered, too, that in adding food – as by digging-in manure or watering the soil with sugar solution or with urine – the change has not been only in the amount of energy-material available; but the soil has been physically disturbed simultaneously by the act of incorporation, wetting, etc. But, given the new set of physical conditions, the microbes best fitted to attack the food seize upon it and increase enormously in numbers. Microbes of one or more kinds thus predominate enormously for a time: either until the food-supply has been exhausted, or until the originally flourishing microbes slow down their own growth through partial exhaustion of the oxygen of the soil air, and accumulation of carbon dioxide and organic acids and other by-products. Then there may be a pause until the air is renewed or some other change in conditions sets in; then, very likely, another set of microbes takes up the tale, and does correspondingly for the new conditions – altered by the presence of live or dead bodies of the microbes which had just before reached their peak, and by the presence of their by-products. Some of these bodies and by-products of the previous generation must be food for the next lot of microbes; and so on in infinite detail and procession: until finally the possible forms of food utilizable by every possible member of the chain in every occurring circumstance have been used up, and the solid residuum becomes 'soil organic matter', or what some people vaguely call humus.

This humus can be of very varying quality. It is therefore not useful to argue that addition of organic matter to soil will at least produce humus; for the soil organic matter depends for its composition and quality upon the conditions which assisted in bringing about its formation: as do the microbes themselves. The quality and productiveness of soil is thus a matter of environmental conditions: it can be changed wisely only from knowing that, and from applying the implications of that knowledge of soil as a seat of balances and equilibria.

Final attainment of the soil-microbial equilibrium after an addition of organic matter or other disturbance of an initial equilibrium may take a long time. If it is attained, and if external conditions are the same as when the first change was imposed, the microbes will be back in the equilibrium with which they started. Assuming that the soil was reasonably well aerated, feeding the microbes will have led to a great outpouring of carbon dioxide and to a formation of water, also to a historical passage of now vanished microbial masses across the scene of the feast, and that is all. If the soil was not well aerated, some undesirable substances may have accumulated and the organic matter may remain more undigested (as in peat and marshes); but ultimately there can be no escape from the operation of natural equilibration.

To sustain microbial numbers and kinds in a balance different from what is in accordance with the natural conditions – and hence in equilibrium with the natural vegetation and the natural level of production of food for man – there must be an artificial, continued, interference: not just by supply of microbial food, but by changing as many of the principal conditions of microbial and plant growth as may be necessary.

The soil microbes are always living up to the limit of their capital. Not only *income*; but the totality of their environment. This is really important, because we depend upon natural and cultivated plants, and they depend upon what the soil microbes and other natural agencies yield to them, plus what we artificially supply in the way of lime and plant nutrients, and therefore upon the sum of what we most artificiously do to alter the conditions of the soil for microbial and plant growth together.

Generally speaking, plants feed on the chemically simplest compounds, such as carbon dioxide, phosphate, sulphate, nitrate or ammonia. They build these up into highly-elaborated stuffs, upon which animals rely. Animals have very limited powers of assembling molecules. For what they do, they must start with already complex groups formed by plants out of their simple ingesta.

Simple stuffs are built up into big organisms via plants. But the big organisms are not immortal. They die. If the elements built up into big (or small) living things were not returned to circulation, all life must come to an end, partly for lack of fresh material, and partly because the ground would be cumbered with dead things. The higher plants cannot make use of one another's tissues, nor of the bodies of animals and microbes, so long as those are alive.

The cycle is completed by the microbes, which are essentially preparers of plant food. They are the chefs of the underworld. Their work is known by unpleasant names – decay, decomposition, rotting, and so forth. Though essential to life, decay or decomposition is looked upon as something foul, because it is often accompanied by unpleasant smells. (The smells usually arise when aeration, or oxygenation, is inadequate; when the wheel of life is turning too slowly, so that we become conscious of the spokes and other details of its structure.) Take de-composition in its literal sense, which happens to be a true sense, and it is no more than the opposite of re-composition, which is surely nothing to be frightened about.

The microbial inhabitants of soil, whether they are actually in soil, or are being carried about in dust, or are living in piles of farm manure or in materials like leaves and animal droppings lying on the soil, are like the house-breakers in a city which is for ever building anew out of the materials of its old outgrown or out-moded dwellings. Human house-breakers do not substantially alter their materials, but microbes re-fashion theirs, and infuse them with the mysteriousness of life once more.

CHAPTER TWO

La Ronde: Air One

Chemists have not always agreed among themselves concerning either the number, or the nature of the principles of bodies. – *A Manual of Chemistry*

EVERY grand design is simple when it is understood. Of the truth of this there can be little doubt, though some purists may object that it would be better to say that every grand design appears simple when it, or its essence, is comprehended. To a musical barbarian like me, a fugue or a symphony sounds very much like a collection of notes with occasional pauses (when I think they have finished), and having different portions in which various numbers and kinds of instruments are assigned roles for reasons which escape me. So might another kind of barbarian look upon Milan Cathedral and obtain a general impression of whiteness and lots of spires, without understanding the simplicity of the crucial plan, to which neither whiteness nor ornamentation is essential. What can be geometrically simpler than two lines at right angles? The pattern of cathedral design is complicated by the introduction of a third dimension, then by pillars and curved apses and other structural features which detract from strict rectilinearity; and by towers and stained glass too. Lacking previous instruction, but given opportunity for travel and comparison, a savage would scarcely be able to deduce in the course of one life that most cathedrals have a basically simple ground plan.

In expounding the beautifully simple essentials of soil microbiology I have the advantage of much previous instruction: which is to say that everyone can benefit intellectually from the work and knowledge of contemporaries and those who have gone before. I need claim no originality except for the exposition. Even in the matter of being able to communicate ideas, I benefit from parents and teachers, and from teachers not my parents. The earliest exposure to literary criticism that I can remember was at Tubbs Road School in Harlesden, Middlesex, when I was about seven. The essay was 'An autobiography of a penny', and I finished it with the simple Anglo-Saxon words 'And then I died.' A teacher pointed out to me that, assuming the existence of a literate penny, it was not rational to imagine it continuing to write after it was dead; and that a tendency to conclude the job to the ultimate must be curbed this side of credibility. Those were not, of course, the words; but (to use ordinary language again) I felt that the criticism was unjustified. Its accuracy took some time to sink in.

Having resurrected this penny is to show that, after all, it never quite died; and it might be taken as a text for this book, if one were in a moralizing mood, and if greater texts were not available. Microbes are potentially immortal, so there is no reason to keep out a lighter tone from a discussion of soil microbiology. If a treatise on soil microbiology keeps to the functions of decay and decomposition and other aspects of morbidity with which the activities of bacteria and other micro-organisms are vitally concerned, it must overlook the other and equally important cheerful side represented by the word *vitally*: for that means not just what is vital to the microbes, but what is vital to them and to all of us.

Soil microbes destroy, and at the same time they create. There could be no large forms of life without the continuing participation of the soil microbes. Without them, there would not be every spring that glad phenomenon of *le réveil de la terre* – creation renewing itself.

All life goes on by means of microbes and water. Having read so far, you may be inclined to think that this is just another specialist's plea for his speciality, or pets. What about the sun, or plants, or oxygen? I grant you the sun (remember the man who said 'I accept the Universe'?; and the comment 'Egad! He'd better!') and the plants. The sun is the source of all energy on earth; or, remembering that much terrestrial heat is produced by radioactive change, it might be more relevant to say that the current income of radiation from the sun is the source of most of the day-to-day energy of living things – though within historic times man has learnt to supplement current income by making drafts upon the solar energy of past time that is stored in fossil combustibles; about which, more later.

Because animals (including man) all depend upon green plants, we can perhaps agree for the time being in saying that all *higher* life depends upon soil microbes and water. This leaves me vulnerable at three points to your critical powers. 'All higher life' is not the same thing as 'all life'. It excludes the microbes themselves. So let me say that in the exercise of their faculty for carrying on with decay and decomposition, the soil microbes do largely take in each other's washing; more exactly, their processes are linked more or less serially, as I have hinted; and, of course, the microbes, like higher green plants, require a certain degree of wetness in their surroundings. For life to continue there must not only be water, but the water must be to some extent free or actually wet: snow or ice, for instance, is not free water and is not wet at all, and so is of no use for any process of growth.

Secondly, you have something if you point to the sea, and indicate that the life of fish or whales does not obviously depend upon soil microbes; granted that some products of soil and of soil microbes are carried into the sea, it might be difficult for me to sustain the thesis that the gambols of a dolphin depend very much upon soil microbes. That would be untenable; so let me say that this book is mainly about soil microbiology, and accordingly I might have said that all higher life on land depends upon soil microbes and water.

However, I do defend the thesis that with one important, microbiological, exception there is no essential difference between the fundamentals of life in the sea and on land (or in rivers and lakes). Chapter Seven goes into that, in some detail. The proposition about life and microbes can meanwhile be put in its most general form: that all life depends upon microbes and water.

This goes much beyond looking upon water merely as a source of moisture or as a support for fish and dolphins. Here I must introduce another personal note to defend myself from a possible charge of one-sidedness, with partiality towards a microbiological speciality. I have not been actively engaged in microbiological work for several years. By training I am a chemist. Thus I don't feel that I have become engrossed in microbiological problems to the exclusion of others; but, having one foot in each of the microbiological and non-biological camps, my interest in making a statement about microbes and water or anything else is just to present the truth as I see it, and here as a contribution to what (for lack of a more specific and up-to-date name) I can call only natural history.

No doubt it is for the reason that botany and plant physiology is a much older science than soil microbiology that the notion has grown up of looking upon photosynthesis by higher plants as the beginning of life. To present the fact that higher green plants (and some microscopic ones) fix carbon from carbon dioxide and release oxygen by means of chlorophyll under the influence of sunlight, is a *convenient* early step to make in introducing biology to beginners and in discussing some anthropocentric problems. It is convenient because plants are familiar objects, easy to talk about, and, if need be, easy to use as demonstration material before a class; and it is easy for a student to carry out a demonstration experiment on photosynthesis for himself. The subject of photosynthesis even by the more tractable green plants and at an elementary level is not without difficulties. It is easy to take a healthy sprig of mint and expose it under water to sunlight, to test the evolved gas, to show that that is oxygen, and to reach the wrong conclusion.

So easy, indeed, are such experiments (including more refined versions in which whole plants were grown in air) that for almost a century and a half the wrong conclusion was taught as a principle of botany, of plant physiology, and of the photosynthetic conditions of life of animals and man on earth. Never, probably, has a false hypothesis been more widely accepted in scientific circles by so many and for so long; certainly, never has any hypothesis been so often the subject of experiment, and of demonstration intended to verify and teach its truth – without its falsity being found out.

In the history of photosynthesis we have a beautiful example of the truth of the proposition that *correlation is not evidence*. To hold that a correlation between two facts is evidence of a connexion between the phenomena represented by those facts is to import a large element of belief, or simple faith, into the argument; in other words: to confuse correlation with evidence is to depart very considerably from the scientific attitude. It is undoubtedly one of the curiosities of scientific philosophy that a simple correlation, established in the first years of the last century, should have been accepted as evidence of a truth for so long (until about twenty years ago) and without much critical examination by experiment or otherwise.

In 1796 the Dutchman Ingenhousz published the results of experiments on photosynthesis in large plants. From these he showed that carbon dioxide and light led to giving-off of oxygen during photosynthesis. By 1804, the Swiss de Saussure had shown that the volume of oxygen evolved during photosynthesis was equal to the volume of carbon dioxide absorbed by the plants. This was a very fine result. It reflected much credit upon de Saussure and his technical ability as an experimenter at a time when the technique of measurement was not generally well developed. The result was a true one, and has been repeatedly confirmed. The more exact the experimental technique, the more nearly the ratio of volumes (or molecules) of oxygen put out and of carbon dioxide absorbed and utilized in higher-plant photosynthesis becomes equal to 1. De Saussure also found evidence for participation of water in photosynthesis, but he thought that assimilation of water was an independent phenomenon of growth.

However, from the equality of the ratios of volumes (or, molecules) of carbon dioxide and oxygen concerned in photosynthesis, it came to be taught that the oxygen evolved originated in carbon dioxide decomposed during photosynthesis.

In modern symbols the equation underlying this hypothesis is crudely

$$CO_2 \xrightarrow[\text{chlorophyll}]{\text{sunlight and}} = O_2 + C$$

the carbon (C) being understood not to be set free as elemental carbon but as being taken up by the plants and somehow converted into organic matter. (This organic matter has been long and widely assumed to be sugar; but that is another story.)

There is no evidence whatever to support the conclusion that the evolved oxygen came from the carbon dioxide of the air. What was known for certain was the equal ratio of volumes of the two gases; and from that the hypothesis or guess expressed by the equation just given has been devised. For a long time this wrong guess held the field as an 'explanation' of the origin of oxygen during photosynthesis, and as a substratum for a great deal of appeal to the importance of carbon dioxide as the source of oxygen for respiration of man and other animals. This view about evolution of oxygen from carbon dioxide during photosynthesis was almost everywhere taught as a fact until the last few years: until 1914 it was not even seriously questioned.

Whether you know all this or not, you will see that on page 23 I have deliberately written something which could be mistaken as putting the old view. You will find the sentence stating that higher green plants (and some microscopic ones) fix carbon from carbon dioxide and release oxygen by means of chlorophyll . . . It almost looks as if it were meant there that the oxygen comes from the carbon dioxide! Modern knowledge is consistent with the view that oxygen released during photosynthesis by green plants (including the microscopic ones) comes exclusively from water.

That puts a radically different complexion on photosynthesis. A splitting-up of *water* to yield oxygen and hydrogen is not only in accordance with experimental evidence: it makes sense in other ways. Most notable is the opportunity given for injecting consistency into a statement of the cycle of life, so as to give a core of sense to our representation of the fundamentals of living. That is no mean achievement.

What happens, then, to the carbon dioxide, if it is not split up in photosynthesis? The modern view is that it is absorbed by the plant as such, with its molecule intact, and is then combined with the hydrogen set free from water. This combination of hydrogen with carbon and oxygen of carbon dioxide is an instance of what chemists call reduction.

In every molecule of carbon dioxide, its carbon is combined with the possible maximum of oxygen: the carbon has been oxidized to the fullest possible extent, so that a molecule of carbon dioxide has no thermal energy, and cannot yield energy by being burnt or by any other oxidation-process: it can undergo no further change through participation of oxygen alone.

Through having been burnt in our boilers and stoves, or in the respiration processes of plants and animals, or in forest fires or other natural processes of oxidation, the carbon present in carbon dioxide cannot advance further along the line of oxidation with release of heat energy. If carbon dioxide is to be of use for the vital processes of organisms, it must retreat: along the line of chemical reduction with hydrogen; and any attachment of hydrogen to the molecule of quondam carbon dioxide must be accompanied by an absorption – which is to mean, an intake – of energy.

The antithesis of oxidation and reduction is a fundament of chemistry. So is the physical contrast of giving out of energy on oxidation and the demand for energy in order that reduction may occur.

We must not assume that carbon dioxide must inevitably

take part in this alternation of oxidation and reduction; unless, of course, we think along such lines as *cogito, ergo sum*. It is not inevitable that carbon dioxide shall be reduced; it might accumulate for almost eternity as a naturally useless material – useless, that is, as a source of carbon and carbon compounds – in the way that vast masses of it have accumulated in rocks and the sea. The amounts of carbon dioxide in chalk and limestone alone are enormously greater than are needed for all forms of life; and to these can be added great quantities in solution under high pressure – virtually in liquid form – in the depths of the sea. But, if we grant the inevitability of life, then carbon dioxide must be reduced; carbon dioxide must get hydrogen from somewhere, and energy also. Photosynthesis is the mechanism by which these things are done.

The obvious source for hydrogen is water. In this book we need not consider in detail other possible sources of hydrogen; it will be enough if we accept the apparent fact that water is the source of hydrogen for all green plants. Photosynthesis is also performed by some kinds of bacteria, which utilize hydrogen sulphide (H_2S) instead of the analogous hydrogen oxide (H_2O) for some purposes, and release free sulphur instead of oxygen in doing so; but these bacteria also require water, and are not exceptions to the most important general rules.

We have, then, in photosynthesis the return of the carbon of carbon dioxide, and also its oxygen, to circulation. This is performed under the impact of energy received directly from the sun. The oxygen is immediately available for breathing and for other preliminaries to oxidation; and the now reduced carbon dioxide is on its way to being able to take part in the life of the green plants and other photosynthesizing organisms, and thereby to fulfil the demands for carbon and other elements, and for energy, of any animals which may use those organisms for food. During life the dejecta, and after death the whole of the tissues, of plants and animals are available for the requirements of microbes for carbon and other elements, and also to furnish the whole of the energy of non-photosynthesizing microbes (that is, of most kinds and the greatest bulk of microbes).

Among the 'other elements' besides carbon we must give special prominence to nitrogen. Supply of nitrogen in suitably combined forms, and notably as protein, is the crux of the world's present problem of food supply. Treatments of the subject of photosynthesis tend to concentrate on the fate of carbon: many of them omit to refer to nitrogen; yet the combination of nitrogen with carbon and other elements is an important part of the immediate sequel to photosynthesis, if it is not a part of photosynthesis itself.

The older discussions of the chemistry of photosynthesis assumed sugar to be the first product formed. This was partly because the writers were under the influence of the idea of the splitting-up of carbon dioxide, and partly because it led to a neat and simple equation it was assumed that the carbon was immediately converted to sugars known to be present in plants. I need not reproduce that misleading equation, and I refrain not just because the fundamental hypothesis was wrong.

It is not yet known what the first product of photosynthesis is: the nature of the first product formed when carbon dioxide is reduced by green plants has not been established. The simplest compound known to be a direct result of the photosynthetic process is a compound of glycerol (glycerine) and phosphoric acid. It contains three carbon atoms, so that some intermediate probably occurs between carbon dioxide and glycerol. Modern knowledge suggests that what is generally understood as a sugar is almost certainly not among the first products of photosynthesis, and thus the old question 'Which sugar is the first product of photosynthesis?' loses meaning.

An early product of photosynthesis – possibly preceding glycerol – may be di-hydroxyacetone $CH_3.C(OH)_2.CH_3$, and this or glycerol might give rise to compounds $C_3H_6O_3$ or $C_4H_8O_4$, among which are substances which are classed by specialist chemists as sugars. Compounds with three or four atoms of carbon combined with oxygen and hydrogen are numerous; they include organic acids; it is therefore possible that acids, or their derivatives, predominate among the very earliest products of photosynthesis: long before a six-carbon complex like $C_6H_{12}O_6$ (typifying glucose and other recognized simple sugars) has been built up.

This apparently discursive and aberrant discussion of the inner chemistry of the green plant during and just after the photosynthetic operation is not as widely off the track of the generality of soil microbes as it may seem at first sight. Its purpose is to 'let in' nitrogen. Since sugars contain only carbon, hydrogen, and oxygen, to discuss them exclusively in relation to photosynthesis is to deprive oneself of a useful and indeed necessary reference to proteins. Sugars can form compounds with nitrogen, and some such compounds are of great importance. The nitrogenous sugar-derivative chitin (*pronounced* kye-tin) forms the greater part of the organic structural constituents of insects, fungi, and some other lowly forms of animal and plant. It is attractive to think that such organisms 'adopted' chitin as an economical way of combining nitrogen at an evolutionary period when protein was at an even greater premium than it now is; but that is by the way.

From what is known about plant chemistry it seems that proteins are formed via a combination of ammonia with simple organic acids. Disregarding the ammonium salts, there are broadly two ways in which ammonia can combine with organic acids so as to get nitrogen into carbonaceous molecules. The series of compounds thus formed are the amides and the amino-acids. Biochemically, these and the ammonium salts can be regarded as interrelated, if one keeps away from the rigid formal treatment given by elementary text-books of chemistry, and has more regard to the high degree of lability undoubtedly existing in the actual chemistry of organisms.

Proteins are known to be largely built up from amino-acids. In the absence of knowledge about the ways whereby nitrogen is combined with carbonaceous fractions of molecules in photosynthesizing organisms, it seems reasonable to suppose that the nitrogen is introduced via ammonia, NH_3, into some simple organic acid with not more than four atoms of carbon. This would lead, sooner or later, to the formation of amino-acids of many kinds.

It will be noted that it is assumed that the nitrogen is already reduced to its fullest extent. Ammonia, NH_3, contains the maximum of hydrogen with which nitrogen can combine. If the nitrogen is supplied as nitrate from a natural source such as nitric acid in rainwater, or nitrate produced by the nitrifying bacteria of soil, or from use of sodium nitrate ($NaNO_3$) or another fertilizer, the nitrate must, on this hypothesis, be converted by the plant from the completely oxidized form of nitrate into the completely reduced form of ammonia. How this is done we do not know; but it will be clear that the hydrogen derived from water put through the photosynthetic process is a strong candidate for the position of direct or indirect donor of hydrogen for the process of making amino-acids and proteins within the green plant, in so far as nitrate is taken up by the plant and used to form protein.

Many, perhaps most, green plants can directly take up ammonia (as such, or as ammonium salts dissolved in water). There is no good evidence that green plants can in general benefit much from absorbing any compound of nitrogen with another element, except ammonia or a nitrate.

ADDENDUM

A modern view of the essential over-all photosynthetic reaction is

$$2H_2O \xrightarrow{\text{chlorophyll} + nh\nu} O_2 + 4H^+ + 4e$$

which I shall only annotate. *nh*ν represents the energy needed, in terms of quanta; *e* represents electrons concerned in chemical change.

The most interesting thing is that neither carbon nor nitrogen is shown. This is because the equation is in the most fundamental form. The equation thus refers to the photosynthetic decomposition of water, with release of oxygen gas (O_2) therefrom. The hydrogen ($4H^+$) is available to reduce carbon dioxide or other substance.

The equation is taken from the article by Dr Sam Granick in *Chemical and Engineering News*, 1953, Vol. **31**, pp. 748–51. Readers interested in a modern presentation of the biochemical relations of chlorophyll and higher life may like to consult it.

CHAPTER THREE

La Ronde: Air Two

> It appears to us much more convenient to adopt the denominations given by M. Macquer to these different orders of compounds. He terms them *compounds of the first order, of the second order,* &c. which leaves no room for any kind of obscurity. – *A Manual of Chemistry*

THERE is, however, another compound of nitrogen which is, from the standpoint of our personal preoccupation with problems of food supply for man and his farm livestock, more prominent than either ammonia or nitrate. Since plants depend so much upon ammonia and nitrates formed in the soil and brought to the soil by rain, it is a metaphysically dangerous comparison to say that any one substance is more important than another; so I had better say that the importance of this other compound of nitrogen becomes weighty as soon as we begin to discuss whether and how we can get enough food to sustain the present or a future human population. Under natural, *i.e.*, wild, conditions the relative importances of the various compounds of nitrogen and of other substances adjust themselves: necessarily, since any temporary disequilibrium is adjusted by deaths or new life until either a new or a similar balance is attained. In recent years man has come so abjectly to disturb the balance of living, and most notably between his own numbers and the products of the soil, that he has made himself hostage to his ability to supply himself with protein – in short, to his ability to provide himself with enough utilizable nitrogen in the first instance.

The third and tremendously important compound of nitrogen as a source of protein is the nitrogen of the air: a compound of nitrogen with itself. Usually called 'free atmospheric nitrogen', air nitrogen is free only in the senses that it is physically unconfined, and is not combined with any other element.[1] Air nitrogen is made up of molecules of nitrogen, each molecule consisting of a pair of nitrogen atoms very tightly combined with each other. So tightly are the members of each pair linked that it requires great expenditure of energy to separate them.

The biggest fact about utilizing air nitrogen for purposes of human nutrition is that, alone among living things, only a few kinds of soil microbes are known to have the power of seizing upon this energetically reluctant air-nitrogen, and putting it upon the road of becoming protein in such a way that we can have protein in concentrated form with a minimum of trouble. It is an odd thing that nothing is known about the mechanism of the first steps in this biological – better, microbiological – fixation of nitrogen. From general biochemical knowledge, and by analogy, it seems a fair guess that the first, or at least the principal product, of biological fixation of nitrogen is ammonia. If that is so, biological fixation of nitrogen is essentially a complete reduction of nitrogen, followed by assimilation of the ammonia and its combination into amino-acids and amides. For lack of specific guidance, we may adopt this 'ammonia' hypothesis.

Biological fixation of nitrogen is independent of photosynthesis. Some kinds of microbes perform it without possessing chlorophyll and without being associated with any green plant. Other nitrogen-fixing microbes are associated with green plants. By far the most important of these microbes, and of all nitrogen-fixing microbes, are the bacteria associated with many species of leguminous plants. These bacteria are called legume-nodule bacteria. They are apparently quite unable to fix nitrogen by themselves, so they need to enter into a special kind of close partnership

1. The atoms of 'free oxygen' in air are typically in pairs, too. That fact is not immediately relevant to the argument about plant nutrition, though it is of interest to note that oxygen atoms must be set free singly in decomposition of water in photosynthesis or otherwise, and must at once assort themselves into pairs.

with their legume host-plants in order to fix appreciable quantities of nitrogen. To that extent only do the legume-nodule bacteria depend upon photosynthesis (*i.e.*, for the existence of their host and partner).

There is a class of microscopic unicellular green plants dwelling in soil and water – the microscopic algae – of which some members can fix nitrogen as well as being able to perform photosynthesis. These minute algae are unique in being able to fix both carbon and nitrogen from atmospheric sources. It may be that in these algae the photosynthetic and nitrogen-fixing processes are linked directly; this appears the more probable if it is remembered that all the vital processes occur within a single cell.

There are some supposedly nitrogen-fixing fungi inhabiting non-leguminous green plants. Heather (*Calluna*) is an example of a plant thoroughly permeated by such a fungus. The outcome of these permeations by a microbe is obscure.

Some non-leguminous plants have developed an association with nitrogen-fixing micro-organisms in root nodules. These nodulated plants include alder (*Alnus*) and bog myrtle (*Myrica gale*), which are typically inhabitants of wet and poorly-aerated places where there is little nitrate or other suitably-combined nitrogen ready-made for plant growth. Another set of inhabitants of wet, and sometimes peaty, situations are the plants such as butterwort (*Pinguicula*) and sundew in Britain, and the pitcher-plants in hot countries. These and other carnivorous plants use various devices to attract and hold insects. When the captured insects have died of exhaustion – either because (in some of these plants) they are stuck to the plant by its glutinous secretion, or because (in other kinds of carnivorous plants) they have failed to find the way out and thus solve the puzzle which the plants present in various ways to intruding insects – the insect bodies undergo a kind of digestion. Carnivorous plants presumably benefit from the nitrogenous, and possibly other, compounds formed during decomposition of the animal matter: provided that they do not get too much of it. An authority has told me that it is quite possible to make a pitcher-plant suffer from a form of indigestion, by dropping an undue amount of meat into the pitcher.

Since the association of legume-nodule bacteria with their appropriate leguminous hosts is the only plant-microbial association known to be important in respect of biological nitrogen-fixation for cropping and production of food, we need consider only the leguminous plants and their special bacteria. These bacteria are often called simply legume bacteria, or nodule bacteria. (The kind of organism in nodules of alder and bog myrtle has not been recognized with certainty, but probably neither is a bacterium.)

For safety I add a few cautions. Some leguminous plants have not developed the habit of associating intimately with any kind of nodule bacteria. About most of the many hundred species of leguminous plants it is not known whether they harbour or need nodule bacteria. The discussion of biological nitrogen-fixation will therefore proceed with virtually exclusive reference to leguminous crop plants (*e.g.*, clovers; lucerne or alfalfa; ground nuts; beans, peas, and other pulses), all of which normally associate with nitrogen-fixing nodule bacteria.

If the nodule bacteria, in association with their host plant, do not call upon it for photosynthetic hydrogen to reduce nitrogen to ammonia, whence does the hydrogen come? Obviously (I feel) also from water; from water of the tissues of the nodules. A more probing question is, however: Do the nodule bacteria release 'free' oxygen when they are decomposing water to reduce nitrogen to ammonia? If not, what happens to the oxygen of the water?

Every molecule of nitrogen, if it is completely reduced to ammonia, requires six atoms of hydrogen; that is equivalent to three molecules of water and to their three atoms of oxygen. This oxygen is not released; so it must be employed in oxidizing something (probably sugar or other carbohydrate, or possibly organic acids). In crop plants the nodules are attached to the roots; they are therefore in the dark, and no energy from sunlight or other external source can reach them. Nothing is known for certain about the source from which the nodule bacteria obtain the energy needed for nitrogen-fixation; but it is a fair guess that it is obtained from a partial oxidation of carbohydrate or organic acid supplied by the host plant. Simultaneously – we go on guessing – most of the ammonia is incorporated into the organic acid (shall we say?) and the amino-acid thus formed is mostly taken up by the host and is potentially or actually available to build protein. Some of the newly-formed nitrogenous compound(s) may be excreted from the nodule or root into the surrounding soil; a small proportion is used by the bacteria themselves for their growth and multiplication.

It is known that a good-going bacterial and legume host-plant association fixes much more nitrogen than the bacteria can use; that is why the host-plant, at least, can become richer in combined nitrogen from a successful symbiosis with nodule bacteria. Besides that, neighbouring plants – and especially non-legumes such as grasses – can benefit from the nitrogen-fixing activity of the legume-bacterial association; for, even though actual excretion of compounds of nitrogen be dismissed as a material factor, sloughing of nodules and fine roots from the host plant represents a form of adding combined nitrogen to the soil in the place where it will do the most good.

We have now dealt with the modes in which the three chief nutritional elements, carbon, hydrogen, nitrogen, together with oxygen, are called down from the air and from water to be of service as food for man and other animals. I am aware that these alone are not enough, and I do not claim to have made more than a start. But this beginning is good enough: for all the other elements in plants and food come from the soil directly and exclusively and are of the earth, earthy.

I am also aware that hydrogen is not generally regarded as a nutrient even when it is in combined form. Nutritionists speak of carbon and nitrogen balances, and also of balances of phosphorus and calcium and other earthy or 'mineral' elements, when they are discussing the income and outgo of the components of food; they also speak of heat balances to correspond, but never of the balance of hydrogen or of oxygen. The reason why no nutritional book-keeping of hydrogen or oxygen can be attempted with profit is that the preponderating mass of both is normally in water used as drink or contained in food.

Though I have perhaps used the wrong word in speaking of hydrogen in food as a nutrient, it is a fair term if we have regard to the energy of food. Animals inhale oxygen, and that is the origin of all the oxidations from which they derive energy for all purposes. They use this oxygen to oxidize, more or less completely, the carbon and hydrogen of the food. Every nutritionally useful organic constituent of ordinary food, except one, contains carbon, hydrogen, and oxygen: that is to say, all constituents of food which can be utilized by man or a farm animal are partly oxidized before they come to be eaten. (The exception is the hydrocarbon, carotene, which contains carbon and hydrogen only; it is the precursor of vitamin A, to which it can be oxidized in the body.) Heat and energy are obtained by animals by carrying on with the oxidation of carbon and hydrogen in food. If the oxidation is completed (as it is to a large extent) the products are carbon dioxide and water.

These animal-induced oxidations are not always complete – for example, vitamin A represents only a slight degree of oxidation of carotene; but so far as it goes, oxidation of carbon and of hydrogen combined with it in the foodstuffs is the animal's own source of energy. Nitrogen combined with carbon in amino-acids and proteins and other organic constituents of food is not oxidized by animals; it stays in the reduced form in which it enters, or is converted to ammonia.

We have thus followed in some detail the familiar pattern of setting output of oxygen and accumulation of carbon by plants antithetically against intake of oxygen and consequent oxidations by the animal body. To that has been added the biological drawdown of nitrogen from the air by microbes. Rain has been mentioned as being the other natural source of combined nitrogen for plants, and hence indirectly for animals (which cannot make any direct use of ammonia or nitrate or indeed of any form of nitrogen except in amino-acids or similar combinations). If 'rain' is broadened to include snow, dew, and other forms of precipitation, we have the outline of a practically complete presentation of the nutritional relationships of plants and animals for everything except the earthy constituents of their respective kinds of nutrients. It has been made plausible that biological fixation of nitrogen by leguminous plants – which is so important for animals in the wild state and on farms – relies upon photosynthesis for its energy.

So we see ourselves and the other animals, great and small, munching and browsing on plants and plant products and transforming them into own bodies and progeny, and excreting a large amount of effete matter on the road through life. Then death comes, and whatever remains to every animal becomes effete.

If that were all, there would certainly be a hiatus. If we consider only the quadrupeds and bipeds of every kind, the hiatus would not be a very big one, for what they utilize or waste in the course of feeding is a quite small proportion of the total mass of dry matter produced anew every year by photosynthesis and its sequential processes. It has been estimated that about 98 per cent of the products of photosynthesis remain as plants. I am not sure whether this estimate was intended to include the sea, but it seems reasonably correct for the land. We may think of the amount of plant tissue which remains on its own feet in the form of trees and other perennial plants, the mass of leaf and timber fall from all kinds of plants, the dead annuals which overwinter, the amount of plant tissue consumed by fire. All such plant material, together with animal material and animal excretions, must be returned to the soil at a rate which approximately equals the rate at which it is formed.

But since green plants cannot eat plants, and cannot, in general, feed on animal bodies, and since the greater part of animal excretions is useless as food for plants: the soil microbes take over the role of decomposing effete substances as far as is necessary. There are some exceptions, obviously; fire does some of the work of combustion (a hint, there); ammonia excreted by animals need not, perhaps, be converted to nitrate. Ruminants pour out into the air an appreciable amount of methane. Methane is fully-reduced carbon, CH_4; a full-grown cow or ox excretes about a quarter of a pound of methane daily, of which three-quarters is carbon; and that amount of methane needs a pound of oxygen to oxidize it completely to carbon dioxide and water. How oxidation of methane is effected is not clear. Large quantities of methane are also produced in swamps and coal-mines; yet there is no sign of accumulation of methane in the air. Probably methane is oxidized in the upper air by ozone to formaldehyde; and if that is brought down by rain to the soil it can easily be oxidized to carbon dioxide and water by bacteria of well-aerated soil. Possibly, methane is completely oxidized by chemical processes.

The global role of soil microbes in returning effete matter of plants and animals to circulation in the forms of the principal plant nutrients and carbon dioxide and water is at last becoming well known. With additions from what is produced by fire on earth and by electrical discharges (and possibly by some direct atmospheric oxidation of methane) the soil microbes complete the cycle of life on land, and even yield a surplus of oxidized products to the sea.

This function of converting effete animal and plant material into food for the next generation of plants is performed collectively, on land and in the seas, by the whole population of microbes of soil and water; in addition, some kinds of soil microbes can fix atmospheric nitrogen. Other kinds of soil microbes make transformations which appear contrary to the cycle of life: some bacteria release nitrogen gas from nitrogenous compounds, and thus go some way towards undoing the work of the nitrogen-fixing microbes. Other kinds of bacteria set sulphur free, although sulphur as such is of no use to green plants; and other kinds transform carbon of organic matter into methane. Such 'contrary' or counter-cycle actions may appear undesirable; but they occur to a notable extent only where oxygen is markedly deficient or lacking. One moral is that good aeration is a condition of good growth of plants and hence of good farming.

CHAPTER FOUR

The Thin Smear

Elementary earth is not so easy to be recognized as the other elements, the principal properties of which we have ascertained. This difficulty arises from the prodigious variety of stones and earths which nature offers us. – *A Manual of Chemistry*

It is not possible to give an exact definition of soil, or of agriculture, or of farming. Leaving aside the very just question whether it is possible to define anything, I mean no more than that soil (or agriculture or farming) can mean many things to different people, so that one talking to another using the same word may fail to get his meaning across unless the ground is cleared by a preliminary explanation.

A French-English dictionary will translate *balai* as broom, and broom as *balai*. This dictionary statement of fact will not prepare people on either side of the English Channel for the other fact that what is called 'a broom' in Britain is a very different object from what is generally understood by 'un balai' in France. The typical British broom is a mostly solid object with shortish fibres projecting from a long cross-piece; the French idea associated with 'un balai' is a thing of long wispy reeds – something like a hybrid between a besom and a carpet-brush, but not very wide, and looking to us, on this side, as an unsatisfying object rather difficult to manipulate. 'Call *that* a *broom*?' we are inclined to ask. (For further notes on the funniness of foreigners, see page 108.)

There are at least three classes of meaning for the word *soil*. Thus, an indexer, or a user of indexes, may easily go wrong unless the word *soil* in the sense of earth, or what the United States citizen usually calls 'dirt', is distinguished from the word *soil* which means all kinds of soilage on clothes and textiles generally, or what a Briton would call 'dirt'.

There is a third meaning of *soil* which is still more easily confused with soil in the agricultural sense. Civil engineers refer to unconsolidated rocks such as gravel, sand, and clay as 'foundation soils' or simply as 'soil'. Since an agricultural soil or subsoil may belong to the same classes (and especially sand or clay) and the topsoil, or soil, may be derived from such materials, the distinction between the agricultural and civil engineering connotations of 'soil' may appear to lack a difference. There is, nevertheless, an essential difference. That difference is well exemplified by the attitude of the builder or civil engineer about to build a house, a bridge, a factory. The first job in construction is usually to remove from the surface what the farmer who tended and cherished that bit of land called the soil.

A civil engineer looks upon that not as something which produces verdure and food, but as something too soft to build upon, and consequently a nuisance. Its only virtue is that it can sometimes be sold at a profit.[1] Having bodily removed what farmers in Britain would call 'soil', the house-builder or civil engineer then gets down either to solid rock or what *he* calls 'soil' – meaning something which professional pride can admit, in default of hard rock, as a suitable support for foundations.

Soil in the detergent sense is thus matter in the wrong place. (Perhaps, from the laundry's point of view, it is in the right place!) In the sense which interests us for the next hour or two, it is the thin smear of material covering the land – often to a depth of only inches, and seldom having a depth of more than a few feet – upon which all life on land primarily depends.

A difficulty at the outset will be to settle: how deep is soil? Should it include the topsoil and all the subsoil – maybe potential 'foundation soil'; if only some of the subsoil – which may be clay or sand many feet deep – how much? There is no rule. Soil scientists usually go no deeper, for surveying and classificatory purposes, than about three feet

1. In Malta it is legally compulsory to put on the fields soil removed from the site of building operations. This is literal soil-conservation.

(one metre). For special investigations – like the one that was not made at Kongwa before the ground-nut scheme was started – it may be necessary to go somewhat deeper, in order to see whether unknown soil has peculiar features (such as very deep roots of the trees, or other evidence of endemic drought or of probable difficulty in bulldozing the trees; and any other features making it improbable that an attempt at growing shallow-rooted annual crops like sunflowers and ground nuts would be a success).

Reconnaissance from an aeroplane or a motor-car, or on foot, is *not* a substitute for having a look underneath the surface. Glory be, and name of commonsense, that's where the soil is: underneath. So, the very first rule in soil science, or in examining the soil for any purpose like taking a farm (whether it has fifty acres or is meant to have fifty thousand or more) is: Dig holes to observe the depth of the (top) soil and any other features the soil and subsoil may present. In any event, your hole-digging will have taught you something.

As a rule, you need not go very deep; sometimes you cannot – if rock or hard gravel lies a little way below. I tell my students that fortunes are never made by digging holes in soil (in the agricultural or horticultural or forestry senses); but small and large losses can be avoided by taking the sensible precaution of digging holes before any money is spent on agricultural development.

The chief interest in soil is naturally in the topsoil. What lies below is important to soil scientists and also to cultivators and potential cultivators. But here I just give a caution. Soil is not necessarily of agricultural interest or importance. The soil sciences include soil microbiology and soil chemistry and soil physics and much else besides. What is called 'soil science' is not any of those named in particular, but is a composite of them all with botany and geology and climatology and so on; even sociology.

Soil science is definable: it is the study of the soil as a natural object. It is comparable with zoology (the study of animals), since in zoology one is not restricted to studying farm animals. Soil science embraces the study of desert soils and tundras and other soils which for any reason produce few or no plants and no crops. It is a historical accident that the teaching of soil science has found its widest outlet and application in agricultural instruction; and, of course, it is in agriculture and horticulture and forestry that soil science finds its most practical applications. It would, however, be a sad mistake to look upon soil science as a branch of agricultural science – in connexion with which the soil sciences are most frequently the subject of teaching. The correct view is that agriculture and horticulture and forestry are branches of applied soil science.

You will find that this is not the only occasion on which those who deal with soil sciences have to complain about the absurd way in which their studies are regarded.

Soil – or, as some people might say, the soil material – is a complex of more or less altered rock on the surface of the earth (or land), and is inhabited by minute living things, and contains the more or less decomposed residues of recent generations of plants and animals. It is usually stratified into layers (technically called *horizons*) of which the topsoil and immediate subsoil are the best-known examples. It usually possesses one or more kinds of structure proper to itself, in addition to any structure possessed by the fragments of organisms and geological minerals it may contain.

Unsatisfactory as it may be, that is the nearest I can get to a definition. The definition is rather unfair to the peats and other 'organic' soils consisting largely of organic remains. It does, however, contain the prime essentials and some of the secondary ones. The two principal points are (1) that soil must contain a population of living things dependent on their environment; these will usually, though not necessarily, include such visible organisms as plant roots, and insects and other small multicellular animals; and (2) that soil is endowed with an inheritance of non-living organic matter as well as living organic matter. By those tokens is soil distinguished from recently comminuted rock: though rock, comminuted or not, if exposed to the atmosphere is potential soil. The reservation about the surface of the earth (or land) excludes marine muds. You may like to argue about that, but I haven't space.

One interesting thing about soil is that it is all on its way to the sea: by transport in solution or as water-borne sediment or as dust.

The most important things about soil, regarded from the point of view of soil as a source of food and as the basis of agriculture and horticulture, also forestry, are not expressed in this definition, but are implicit in it.

The most valuable service I can do for you, in talking to you about soil, and assuming that you are interested in food (not as gastronome but as a student of the processes whereby food comes from soil) is to lead you away from the ancient preoccupation with its mineral solids. You have most likely encountered some book of geology or physiography or natural science which has treated of soil and has put the inorganic solids in the forefront, and has probably made a point of their value as plant nutrients, and thus to agriculture. This view will probably have been temporarily reinforced by my references to altered rock. So, lest you be confirmed in believing something which is hardly more than a half-truth for most of the agricultural soils of many civilized peoples of to-day, will you let me induce you to throw away those failing geological crutches about the always preponderating importance of the inorganic solids of soil?

I imply that that view is old-fashioned, not only because there is a newer, more plastic, and dynamic view, nor because of the fact that living and non-living organic constituents of soil may be as important as, or more important than, the inorganic solids derived from the rock parents of the soil material; but also for the very practical reason that it does not fit the circumstances of modern, intensive, farming.

For a while, however, we can forget special problems of farming, and consider some important generalities about soil – as a natural object or as a cultivated and therefore to some extent a non-natural object.

As will be seen from the attempted definition, the emphasis of the soil is on life. The soil harbours life, and contains an essential proportion of the products of previous life. It is also the basis of above-ground life on land; but the whole assumption on which this book is written is that *that* cannot be fully understood unless the essentially biological character of the soil is first made as plain as he can manage (by one of us) and understood (by the other partner). So, we can revert temporarily to considering the soil itself.

To say that the emphasis of the soil is on life is not quite the same thing as saying that the soil is living. Some people who are not scientists like to pretend that the soil itself is alive. That is a pretty metaphor, and is almost true. It is still more nearly true to think of the soil as an organism. Strictly, of course, to call the soil an organism is to imply that it has a life of its own, as distinct from a population of living things contained in it and making up a large and indispensable part of it.

The idea of the soil as an organism capable of having its chemical changes investigated in much the same way as biochemical changes are studied in a living mouse or an excised organ was indicated by Professor J. H. Quastel when he introduced the term 'soil metabolism'. Metabolism means all the chemical changes within an organ or organism that are associated with its living processes. Metabolic changes in soil include respiration, formation of nitrates and ammonia (pages 135–6, 140–4), and many others; they are, of course, due to the soil bacteria and other microbes. Strictly speaking, it is the metabolism of soil material that is studied, since the investigation is made on a small sample – not on the complex of horizons which constitute most natural soils.

If you wish to think of the soil as an organism, I would suggest that it might be compared to a coral reef. But if you will not go so far as to endow the soil itself with life, I would suggest that it might be better called an organization than an organism.

It is an exceedingly flexible – almost infinitely flexible – organization. With a given kind of parent rock-material, the soil develops a set of characters and of inhabitants (microbial and plant) which are exactly in accord with the circumstances of situation: where man does not too explicitly interfere. Those governing circumstances are largely climate and relief. I hardly think this doctrine of perfect flexibility and exact accordance requires proof; I take it to be part of the nature of things – though later on there is an attempt to portray something of what I feel must be. (My feelings on the matter are aided by what I know generally and specially about soil and water.)

As an example of something you will already know, and can confirm further if you care to take the trouble, will you think of a poorly-drained field of rough grass and rushes near the edge of a lake, and level enough to be often flooded. There are always wee hummocks and patches of varying size a little – perhaps a few inches – above the general level. It will be clear that these will be less wet, or wetted thoroughly less often, than the rest; therefore they will be more, or more often, exposed to oxygen and oxidation. Whether they have a different microbial flora you may not be able to detect; but do they not bear a different – if only a partly different – kind of plant from the generality? And if you dig into the hummock with a trowel, will you not find that the substance or soil of the hummock has a different structure (and perhaps a different texture) from that of the level ground just a few inches below? That will be an example of the influence of a micro-climate; but if the soil and vegetation adjust themselves so closely to such minute difference in environment, will you not readily grant the rest – including the deduction that the soil's micro-population must have altered very considerably too?

The notion of the soil as an organization would make the soil quite comparable to a factory, being a series of workshops. And that is precisely what the soil is.

Let us leave (to Chapter Five) questions of workshop organization and of what makes the factory tick, so as to confine ourselves briefly now to what is done in the soil. We need to discuss briefly the raw materials and the end-products: the latter being taken to be as it were at the factory gate, so that we do not have to concern ourselves much with plant physiology or other aspects of problems of what plants do with what is put at their disposal by the soil.

The point that will be obvious to-day is the one which Lavoisier (*see* page 213) and others long ago could only guess at: that animalcules or microbes in the soil constitute the missing link between the life of plants and animals and the mineral world. The link, however, is not primarily between the three kingdoms vegetable, animal, mineral. (Any full discussion must, of course, link them all; but to think initially in those terms would confuse the issues by making them too complicated for an introduction; besides which, 'minerals' mean different things to different people.)

The soil microbes do play a big part in helping to release plant nutrients from soil minerals in the geological sense and other inorganic solids of the soil. Nevertheless, the microbes do that very largely because of the carbon dioxide they produce during decomposition of the organic remains of previous life. Furthermore, since animals depend upon plants, it will be sufficient if I treat for a while not of the plurivalent cycle of 'the three kingdoms' but simply of the cycle: plant to plant. If our interest were mainly in farming, it could be said that we have to consider the next crop, or at most, the next few crops: say, one rotation.

We have seen that the most notable chemical actions done by plants *as plants* are in relation to two gases of the air: oxygen and carbon dioxide. The air will be moist, if only because of the moisture given off by plants. Still, there is so much other moisture available that we can dismiss the water-vapour of the air with the remark that those early students of photosynthesis seem to have overlooked the need for any form of water in that detail of plant chemistry. In other respects:

Water is essential to life; it is not just the menstruum wherein life goes on.

Water has a much more extensive role than that of mere

solvent in blood or lymph and plant sap and the sea and the soil.

Water is the very medium (in the literal sense) or median and centrum of life.

Water plays physical or mechanical roles in the sea and in the soil: as support, and as a medium of circulatory exchange; it has a physico-chemical role as solvent; and it has a chemical role as provider of hydrogen and oxygen for all the processes of the chemistry of living.

Life consists of a set of chemical processes specialized and cunningly disequilibrated for special purposes of each species, and of each organ in the multicellular organisms; but over-all it is and must be an exact balance of oxidation and its converse, reduction. Life in sum accords exactly with the exact chemical equivalence of the hydrogen and oxygen in water, leaving nothing over. There are fluctuations in the amount of water that is vitally employed; it is always an infinitesimal fraction of the water on the earth; there are variations in the way water is used for purposes of living; but, while it is true that man is dust and shall return to dust, it is water that is the *spiritus vitalis* for the two opposites, the *yin* and *yang*, the alpha and omega, the hydrogen and oxygen, the matched electrons; and so water is both the lubricant and the stuff of living.

Clearly, since the plants decompose water to provide hydrogen for use in their own immediate purposes of utilization of carbon dioxide, and to make oxygen which is set free: then, wherever the soil microbes are in contact with air and its free oxygen they must collectively use oxygen to produce something which plants can employ to balance the hydrogen newly taken up within them.

That is the grand design. Life goes on in accordance with it. We can see and say, at the risk of an anti-climax, that besides that omnipresent compulsion, there is the conclusion that, if that design did not exist, chemistry would have no meaning.

The chemistry of life makes sense if it is understood that the cycle of life can be put into the form:

$$\text{water} \xrightarrow{\text{green plants}} \text{(hydrogen)} + \text{oxygen}$$

$$\text{oxygen} + \text{plant remains} \xrightarrow{\text{soil microbes}} \text{green plants} + \text{water}$$

On this view, there is no necessity for the larger animals, including ourselves. They come in as a complication. But – if only because there are protozoa and many other kinds of microscopic and other small animals (such as mites, insects, myriapods, earthworms) of which some must, and others may, be included among the soil microbes; and because it is impossible to say where the dividing line should come between animals small and large – a few notes about animals will not be amiss.

Animals, like plants, respire; that is, they take in oxygen for use in oxidation of carbon and hydrogen within their tissues, and form corresponding amounts of carbon dioxide and water. So do most kinds of micro-organisms. Animals, unlike plants, are purely oxidative organisms as far as their relationship to the atmosphere is concerned. Plants are both users and producers of oxygen. With plants, the nett amount of oxygen produced is much greater than the amount of free oxygen consumed. The vital point about plants is that – alone among living things – green plants are able to produce oxygen from water, and can use the hydrogen which is thus left for an instant partnerless.

It is also important that plants are able to employ hydrogen from water so as to reduce nitrate (from the soil) and to build up its nitrogen into proteins. This may be an exclusive ability of plants (not necessarily of green plants). Some bacteria can reduce nitrate, but no animal can.

There are many possible annotations to the uppermost theme of splitting of water by green plants under the influence of sunlight. It is a condition of all life in the sea as on land. The relative amounts of water in the sea and on land should not be allowed to be a distraction; in any case, the amount of water involved in the annual turnover of oxygen for respiration everywhere and of hydrogen for green-plant synthesis is small in relation to the total of water; consequently, the amount of water produced by animals, plants, and microbes in their oxidative capacities and processes is also small, and may be regarded as quantitatively negligible except in special studies.

On the matter of quantitative balances, there is the fact that much carbon dioxide is released by volcanic action, and further amounts are produced by other agencies than microbes, so that it might appear that there is not an equilibrium of carbon dioxide. In recent years man has burnt increasing amounts of fossil carbon; these quantities are not negligible even though they are small in relation to total amounts of carbon dioxide in the air. In view of man's industrial output of carbon dioxide, it is sometimes asked whether there should not be a greater profusion of vegetation to correspond. There may be a slightly greater growth of plants as a result of use of coal and petroleum by man. My concern is with the essentials of living, and not with quantities; so I will not discuss in detail the quantitative book-keeping of carbon dioxide. It will suffice to mention that in such matters of equilibrium between atmosphere and earth, the sea (considered physically as a solvent, and not as a medium of life) is a vastly greater equilibrator than are the biotic actions occurring in the sea or on land; so the question about increased growth of plants is of secondary import.

* * *

THOSE who have considered only oxygen, or oxygen in relation to carbon dioxide or photosynthesis in general, must be held to have overlooked the other biotically necessary gas of the air. To give a pertinent and useful indication of the serious omission to take account of nitrogen is to do much more than go over familiar ground. It may not be unnecessary, nevertheless, to recall that animals cannot make the materials of their own protein.

If animals are given the initial materials from which they can start, they can build up their own forms of protein. These starting-points or building-bricks of protein are the amino-acids – compounds containing carbon, hydrogen, oxygen, nitrogen, and sometimes sulphur. In plants, animals and microbes, the amino-acids exist partly free, but are mostly combined into the organism's protein. There is, however, *no* source of amino-acids except what plants and microbes can build up from nitrate or ammonia and the carbonaceous products of photosynthesis; or what a certain few kinds of microbes can make from air-nitrogen if they are supplied with carbonaceous materials.

Thus, whether one is a vegetarian or not, nitrogen for animals is the core of the food problem. It would be very unhelpful to discuss photosynthesis, or the cycle of life, or the roles of microbes in the soil, or any other aspect of the thin smear upon which our life depends, without bringing in nitrogen as the essential partner of the team of elements directly involved in photosynthesis.

Though, in general, nitrogen does not appear to be directly concerned in the photosynthetic process, the fate of nitrogen in organisms is bound up with green plants, and hence with the cycle of oxidation and reduction being maintained above and below the surface of the soil, and straddling that surface through the agency of oxygen and hydrogen derived from water concerned in photosynthesis.

That cycle will go on – we can feel sure – whatever the fate of man. It would not be very percipient to ask at this stage: 'So what? Why bother, if it's bound to go on anyhow?'

The food problem for man no longer depends solely upon what man can get out of the thin smear of soil. Once – and not very long ago – it did depend on that alone, plus some fish. Man's food problem is still a question of what he can get out of the thin smear, plus some fish. But there are now many more people to feed than there were when man depended in principle on the thin smear, and its processes, alone.

The food problem is now a question of activating the processes of the thin smear and all that it implies; of pepping them up, if you care to say so. The problem cannot be solved in terms of greater appeal to photosynthesis alone by putting together more carbon and hydrogen and oxygen so as to make more starches and sugars, with an incidental accompaniment of proteins built up by ordinary plants; it would be still further from solution if we managed to perform carbonaceous photosynthesis artificially. Either result would leave a void; the nitrogen problem would be unsolved, or aggravated. Man cannot live by bread and sweets alone; no animal can.

From the food-production point of view, the cycle of life must be regarded as a producer of carbon and of nitrogen in forms assimilable by man and in balanced amounts – amounts, that is, which correspond to the nutritional needs of man, and of his animals in so far as he is not a vegetarian.

The cycle revolves only in one direction. If the cycle is considered in a totally detached manner, it will be seen that it must go on so long as there is sunlight and green plants and microbes: in particular, it will not matter what proportion of the nitrogen is contributed by atmospheric electrical discharges or by nitrogen-fixation in modern factories, and what proportion by the unique nitrogen-fixing micro-organisms of the soil. Whatever those proportions may be, plants can somehow utilize the nitrogen combined by those agencies. The result upon microbes and plants, and consequently upon the animals which inhabit the earth, will be always and simply that a new balance, or set of balances, will be established rapidly to accord absolutely with the supplies of nitrogen. In that balance, account will be taken of the forms of nitrogen supplied as well as of the quantities.

This proposition is irrefragable, though it is impossible to say in advance what natural balance between plants, or between plants and animals, will be established for any particular conditions of nitrogen supply and sources of combined nitrogen.

If, however, we cease to argue on a plane of irrefutability, and come down, instead, to earth so as to ask 'How does this affect mankind?' we require to know more than that a cycle exists. We must know more about the soil, about the functions of microbes, about plant growth, and about man's nutritional needs: in short about nutritional requirements for animals and about the methods of filling those needs by obtaining food.

After that, it will be possible to discuss whether supply of food can be made equal to some future demand. Please note the *whether*; not *how*.

There is not much difficulty about the *how*; in pure theory or in conversation, that is. About the general principles enough is known for almost anyone to produce a plan or plans for feeding almost any number of people. The broad, casual – almost, it seems, automatic – outlines will be familiar. Use of more fertilizers, of more irrigation, higher-yielding varieties of crops, better breeds of livestock; reclaiming the desert, or farming the sea, and so on – including schemes based on artificial photosynthesis and the like. Many people would sum these ideas up into some such words as greater reliance on science, or taking more advantage of what science offers and is likely to offer.

All will be aided somehow by the white hope of (a)tomic energy, which will carry on for our descendants when supplies of coal and other fuels are exhausted . . .

However, nobody has yet brought out a scheme for converting tomic energy into human or animal energy.

What we use to carry out the vital processes of our bodies is chemical energy. Energy derived from oxidation, for example the energy we derive from oxidizing food, is not always replaceable by energy of falling water or energy from the splitting of atomic nuclei. For animals, including ourselves, such replacement or exchange cannot be made so as to allow life to go on.

The contribution which external energy can foreseeable make to food-supply is essentially indirect. In proportion as adoption of tomic energy or hydroelectric power spares fuel (to say nothing here about the consumption of metallic ores) so will it make fuel disposable for food-production and other purposes.

One great and simple truth to be remembered by planners and their critics is, that however spectacular and long-lasting a source of energy may appear to be, no amount of external energy from any source will be of the slightest use to us, now or in the future, unless we also and simultaneously have something suitable to put between our teeth.

That goes for the sun also.

So the basic question is *whether* enough to eat can be assured. In order to answer that question intelligently, further examination of the principles of food-production is desirable.

CHAPTER FIVE

Soil from the Grower's Point of View[1] (with a Section on Food Preserving)

> The Chemists, in examining a great number of stones and earths, remarked that there were many which had similar properties, and others which differed essentially among themselves. This led them to establish several species of stones and earths. The Naturalists, too, have given their divisions; but as these do not always agree with chemical experiments, we shall not particularize them. – *A Manual of Chemistry*

THE fact that soil has a granular texture has led to numerous misapprehensions from which many teachers (and consequently, students) are still suffering. Mistakes arising from the confusing of soil material with soil, and from looking upon soil as primarily a system of grains, are firmly established in what passes for knowledge about soil. Not only agricultural and horticultural education, but the practices of agriculture and horticulture, have suffered from errors induced by neglect to take account of the vast importance of the spaces between the grains, and of what is to be found within those interspaces.

There are many granular materials commercially valuable because of the composition and quality of their grains; for instance, malting barley, coal, and face-powder. Soil is evidently different from these – not just because it is not usually an article of commerce by weight or in packets, but because it is grains plus.

Suppose you wish lump or powdered coal as fuel. In order

1. Pages 56–65 are condensed from an article 'Something about soil and soil water from the grower's point of view', in *Growers' Digest* (official journal of the Scottish Fruitgrowers' Research Association) 1949, 1, No. 2, pp. 17–27. Thanks are given to the Association for kindly permitting use of the material here.

to judge of its performance, it is sufficient to know the chemical composition of the grains, and their size-range. For some granular materials, other qualities besides composition of the solids are important: germination-percentage in seeds, or, for face-powder, the sizes and shapes of the particles. If such physico-chemical data relating to the solids are known, the rest follows, for most materials consisting principally of solids. Not so with soil.

The amount of water in soil is a historical accident, an incident of the moment. To know the percentage of water in soil is not in the least useful to a grower. That is an illustration of the profound philosophical principle that no fact is of any use by itself.

There is no purpose or point in getting a fact, or a collection of facts, unless it can be placed meaningfully in relation to another fact or facts. The increase in the ability of scientists to relate complex numerical facts cogently and usefully, so as to give a reasonable basis for action, or to say rationally that no clear directive for action exists, is the great new philosophical gain from modern statistical method.

By knowing the percentage of water in coal, it is immediately possible to make an estimate of the proportion of solids. Knowing the percentage of water in field or garden soil gives no information about the proportion of air which is equally important with the water; nor does it tell anything whatever about the amount or proportion of water that is available for growth of plants: not even if a fair proportion of air is assumed.

The difference between soil and other natural or artificial objects made up of grains lies deeper than is suggested by a comparison of dry materials with damp or wet ones. By knowing the percentage of water in, say, coal, waste lime, or sewage sludge, you assess a rather useless and static material (water) which comes along with the solids.

The water and air which are normal to soil are dynamic parts of it. So much is that so, that their movement is of the greatest importance. Water and air in soil must be regarded as floats or fluxes, and should not be regarded as percentages of actual materials as if they were stores for consumption, or as something detracting from valuable solids. The soil cannot be profitably studied as a mass of solids; nor can plants and microbes grow and operate unless water and air are in passage.

Also, soil needs to be distinguished from soil material, even though water and air be included among the latter when it is uplifted.

Natural soil is a part of the landscape; but it affects, and is affected by, the vegetation. Consequently, in wild nature – upon which wild man has little effect – the soil and vegetation, the landscape and climate, are but different expressions of the soil-forming processes.

Nevertheless, man comes into the picture as a potent agent in forming and modifying, and sometimes destroying, soils; and all man's works, including field and garden and glasshouse soils, are fair subjects for detached and rational study.

Soil possesses grains and other solid matter. Of the inorganic matter, some consists of more or less altered rock; but some of the (inorganic) colloidal minerals are apparently formed within the soil and are peculiar to it. Notable among the solids are the organic colloids, loosely and collectively sometimes called 'humus', which are not granular as a rule.

In soil, however, the particles – granular or colloidal – are relatively unimportant. So far are the soil's various kinds of particulate solid matter – inorganic and organic, living and non-living – unimportant that they can be dispensed with entirely. Good growth and cropping can be obtained without them, if certain conditions are observed: as in water-culture (page 61). It should be remembered that water-culture is also air-culture; since the solution of nutrients, which is the basis of water-culture, requires to be profusely aerated.

Water-culture is a method of supplying support, nutrients, water, and air by wholly artificial means. The success of management of cropped soil depends on the extent to which these same needs are met. Whether the means adopted by the grower who uses soil are less or more natural or artificial is a question of the degree of the intensiveness of his cropping; that is, whether he is rancher, farmer, market-gardener, or water-culturist, or something in between.

The idea of soil as a collection of substantially little more than grains and colloids needs to be replaced by the notion that soil is essentially a *workshop* in which nutrients are prepared and dissolved for assimilation by plants.

The larger particles of soil provide not much else than a framework, or fabric of the building; the inorganic and organic colloids may be looked upon as the work-benches with small cupboards and drawers for immediate reference; and – naturally enough – the work of the soil is done in the spaces, on and around the solids.

That is the general picture. For leguminous plants we need to imagine the nitrogen-fixing nodules on the legume plant's roots as a sort of annexe withdrawn from the general factory, and performing a highly specialized purpose while remaining dependent on the general apparatus and systems of ventilation, transport, and other internal amenities of the larger factory; and thus doing no more than helping to deliver the chief quality-products through the main gate which leads to the consumers: us.

Plants can build up their tissues only from such chemically simple (and for the most part, fully oxidized) substances as water and carbon dioxide and the nitrates, sulphates, phosphates, and chlorides of common base-forming metals like calcium, magnesium, potassium, and manganese; also ammonia as such, or as ammonium. Some of these substances would be scarce or absent in soil if the supply were not continually renewed through the combined work of bacteria, fungi, and small animals of soil. This population of soil inhabitants takes over the chemically complex remains of animals and plants, and breaks them down and oxidizes them into substances which plants can use.

This preparation of plant nutrients by soil micro-organisms goes on in and by the presence of both water and air; and solution of the prepared nutrients occurs simultaneously. Under natural conditions the two functions of preparation and solution are not separated. However, in intensive agriculture and in horticulture the soil's function of preparation of plant food is partly replaced and supplemented by addition of fertilizers. So it is logical and convenient to discuss separately the soil's tasks of preparation and solution.

Preparation of plant nutrients cannot proceed without air in the soil, if only because air is the main source of oxygen in a healthy soil. (Some dissolved oxygen and other materials are added by rain.) In partial or complete absence of air there will be a shift of microbial population away from the predominance of kinds which operate in a well-aerated soil, towards those which produce substances useless or toxic to plants – such as organic acids, free nitrogen, nitrites, or sulphuretted hydrogen. Nor can the roots assimilate without air to supply oxygen (also the nitrogen needed by leguminous plants) and to take away the respired carbon dioxide.

To keep the cultivated plants fit there is a need for a continuous transfer of solutions and gases. This need is met by the soil water and the soil air. The incoming and outgoing water acts as a pump; the soil water and soil air jointly constitute a transport system, in addition to performing their other, largely chemical, functions. Our workshop analogy must therefore be extended to comprise a series of workshops having a complicated conveyor system.

Thus, soil is much more than soil material. The difference is at least as meaningful as the difference between a tramcar and a transport organization. Just as a knowledge of electrical engineering or of vehicle-building is of small account on the operational side of passenger transport, so a knowledge of geology or of chemistry is insufficient for an appreciation of processes in the soil.

What we call 'soil' when we examine a sample on a spade or in a laboratory analysis is not really soil; it is soil material taken out of its normal surroundings where it was in contact with, and at least potentially able to have exchanges with, the water and air above, below, and on each side. A soil sample has had its conveyor system disrupted; it is the workshop at a virtual standstill.

Soil may be regarded as having these six main functions:

(1) As a support for plants and animals (including man).

(2) As a long-term reserve of plant nutrients. This function appertains to the interiors of the grains, of which the contents may not be released for centuries.

(3) As a storehouse of a small but renewable reserve of plant nutrients on and between the grains, for consumption by the current crop and the next few crops.

(4) As a reservoir of water, of which normally a large proportion exists as a film over or in contact with the colloids. Whatever the state of this water, *much or most of it is of no use unless added to from outside.*

(5) As a nidus for the activities of micro-organisms and small living things of all sizes from the microscopic up to, say, earthworms. All these co-operate to prepare plant food for the current crop from the last generation of plants and animals (though with a certain carry-over which is the basis of soil fertility in the sense of sustained productivity).

(6) As a means for holding nutrient solution and air, in such a manner that both water and air are available to the plant, and can be easily *renewed*; *i.e.*, can be exchanged with each other and with outside supplies.

None of these functions is indispensable. Life goes on in the sea without (1) as it exists in soil, and without (2) and (4). In water-culture, only (6) operates in the medium which is in direct contact with roots.

For the mere continuance of life in the natural cycle, the most important of the above-mentioned functions are (5) and (6), both of which go on, necessarily, in the pores or spaces between the soil grains. In nature (5) is very important indeed, since it is the chief means by which nutrients are made available to plants. It is supplemented by (2), and some outside agencies, yielding plant nutrients by mainly chemical means. In tropical rain-forests the bulk of the plant-nutrient capital is locked up in the trees and other vegetation, there being often very little in the soil itself. As soon as a tree falls it is actively attacked by microbes (including termites) so that decomposition, and return to circulation, proceed largely above the surface of the soil proper.

Normally, however, most of the current needs of natural vegetation for nutrients are supplied from transformations wrought by biological agencies in the soil itself (5); and a small and partly colloidal reserve (3) is often built up from the remains of living things (also from 'insoluble' fertilizers), and is deposited between the grains.

These activities (3, 4, 5) of the soil's micro-population in preparing plant nutrients, along with function (6) of putting the nutrients effectively at the disposal of plant roots, constitute a large part of the natural history of soil. They may be called the space-dynamics of soil. It will be seen that the functions of that part of the soil's solid matter that is usually regarded as coming within the scope of geology have dwindled to (1) and (2). These are passive; and, while they determine *natural* equilibria, (2) is not good enough for farming and cultivation in a modern signification.

A detailed geological and mineralogical prelude has been customary as an approach to elementary study of cropped soils. Such an approach is a legacy from early mechanistic and chemical views of soils; and the soil's mineralogy must be included in any full discussion of soil as a natural object. Yet it seems doubtful whether conventional ideas about the value of geology as an introduction to study of the relation between soils and intensively-grown crops can be sustained.

Knowledge about the inorganic solids is of great importance in the beginnings of new cultivation overseas, in soils either virgin or in such situations that it is impracticable to supplement the natural output or release of nutrients from the solids; there, not only the amount of nutrients released, but the *rate* of release of one or several of them, may be a decisive factor in the enterprise of cropping. This outlook upon the extent and rate of release of plant nutrients, however, is an appurtenance of the modern science of soil chemistry rather than of geology. The geological sciences are old, with a respectable history which began long before soil science and agronomy or crop husbandry took shape as entities; and attention to such a superficial matter as soil has always been a side-issue for the geological muse. It is not surprising, therefore, that the geological outlook upon soil has not in general advanced as rapidly as agriculture has.

Roughly speaking, the more intensive the production from the soil, the less important the consideration that need be given to the nature of the inorganic solids: always provided that the right sort of space-relationships initially exist or can be induced to form. Thus, in glasshouse work in the United Kingdom, the soil is chiefly valuable for its mechanical function of support, and as a medium through which plant nutrients, air, and water are put rather artificially at the disposal of the crop. The nutrient function of the soil itself is largely ignored – though the microbial-workshop function remains important for adequate transformation of organic manures and other fertilizers given by the grower.

By applying fertilizers, organic and inorganic, the grower can, and does, largely short-circuit the natural production of plant foods; and at the same time he augments the supply of plant food to a level well above that natural to the soil.

A high plane of plant nutrition brings responsibilities: as will be seen later. However, few of a farmer's or grower's troubles about plant nutrition are soil troubles; the majority are solution troubles occurring among the soil grains, but otherwise having little to do with them directly.

The famous definition of a house fits the soil exactly: a machine for living in. The inhabitants are plant roots and soil microbes. Except to an architectural student or a timber merchant, it may matter little whether a house is built of brick or stone or wood, so long as it is adequate for its purpose. To those who live and work in it, the most important thing is not the material of the walls, but the greater or less size or pokiness of the rooms, coupled with the way according to which those essential spaces are more or less conveniently arranged, connected, and equipped. The factory analogy of well-aerated soil is but an extension of the house idea, to include the notion of a conveyor-belt or other device for facilitating exchanges of materials within the building and with the outside.

Soil 'water' is not pure water, but is a solution of many things, including gases; and the air in the soil is not of quite the same composition as the air just above the plant canopy. The composition of soil 'water', or 'soil solution', is modified by the grower's use of lime and fertilizers and organic matter; also by the demands of the growing plant, by the colloidal and other properties of the soil itself, and by many other factors – including day and night.

Nothing in cropped soil is constant for long. Soil is a complex variable equilibrium between solid matter (inorganic and organic, soluble and insoluble, alive and not), water, and gases; its status is affected by interchanges with the atmosphere, and the whole is complicated by presence of growing plants. Altogether, contemplation of what may be the true picture is enough to daunt anyone.

We have made our bow *pro tem.* to the soil micro-organisms and to the powerful organic colloids they yield; and we can assume the presence of roots of a growing crop. For practical purposes it will therefore be sufficient to consider the other occupiers of the spaces between the soil grains; namely, the air and water – or, rather, the solution of whatever is available to be dissolved. It will not influence us, then, whether the various kinds of dissolved matter are gaseous, or have been formed by chemical action on the rock particles, or produced by living things inhabiting the soil, or are supplied out of a bag by the grower. Water and air must both be present *and renewable*, or there will be no crop.

It will be seen that we have got away from the soil grains to the substances between the grains, as being those (subject, of course, to the fable about the stomach and the members) most vital to the crop. This is the modern view. It may be summed up by saying that soil should be looked upon as a cellular rather than a granular material; that the spaces between the grains are far more important than the grains themselves; and, a little more crudely, that what is of greatest importance in soil is what is not there – meaning by that the spaces which can be occupied for good or ill by air and soil solution.

Except in the heaviest types of clay soils and some other special situations, the soil air will largely look after itself if the water conditions are right. Water in excess is harmful because it wastes heat; also because it may connote cloudiness. In regions like the British Isles, a surplus of rain on the soil probably does less to restrict growing of arable crops than does a surplus of cloud. Excessive natural water (a high water-table) can and should be detected by the simple means of digging holes and interpreting what is seen.

When a farmer drains land, he believes he is doing so to get rid of excess water. So he is; yet drainage is basically but a step towards better aeration. Other amelioration results, or can be brought about, after the excess water has gone from the soil. As it goes, it sucks in fresh air, and the ratios of water to oxygen and nitrogen and carbon dioxide are improved totally and severally.

Being neutral towards man, the soil microbes can – and do – adapt themselves exactly in numbers and kinds to any conditions. But if we selfishly leave out of account the anaerobic bacteria which predominate in airless places such as marshes and the bottom of the Black Sea[1] (you will doubtless have smelt their products sometimes; perhaps while digging a little deeper than usual in the sand near high-tide mark) we can look upon the crop-productive, oxidizing microbes as able to get on with their primary job of oxidation of effete matter, once they are supplied with a situation in which they get enough air frequently changed. The conditions of aeration suitable for the generality of

1. That Sea contains as much sulphur, bacterially produced as hydrogen sulphide, as would meet all the world's needs for sulphur for a century at least.

oxidizing microbes are also right for the protein-producing activities of nitrogen-fixing legume nodule bacteria.

If the air-pump sticks, so do oxidation and nitrogen-fixation slow down or stop. Life does not then stop in the soil; but it changes its character and intentions, with the result that the cycle becomes kinked, and slows down appreciably: even to the point where no green plant will grow.

So everything depends, for us, upon the soil spaces, and what is in them and passing through them.

FOOD PRESERVING

FOOD-PRESERVATION implies that something – usually kinds of microbes – shall not grow. Like growing, food-preservation relies on control of water. Whereas agriculture depends on control of water to give the best possible growth under the prevailing circumstances, food-preservation controls and manages water for just the opposite purpose.

Preservation of food and drink by human agency relies on one or both of two broad methods (besides the very special techniques of aseptic collection, and employment of chemical preservatives such as sulphur dioxide, neither of which will be discussed here).

(1) The microbes able to spoil food through their growth and proliferation are removed bodily by filtration, or are rendered inoperative by killing or inactivating them – usually by heating.

(2) Water indispensable for growth is temporarily put out of the effective grasp of microbes by drying or freezing the foodstuff, or by using other means to increase the concentration of salt, sugar, and/or acid in the water which the food contains.

The first method is effective so long as stray microbes, say from the air, have no access. The water in the food is still available for growth, and the food must be protected from entry of microbes: contamination. This method is therefore associated with bottling and canning of products meant to be consumed shortly after the container is opened. Typical examples are unfermented apple juice; evaporated milk; sterilized milk.

Diverse as the second group of techniques may appear to be, they all depend on one principle. This is, that if the free energy of water is lessened sufficiently, the water which remains becomes unavailable for purposes of growth. If soil is dry, there is not much water there, and what is present is largely unfree; but give sea-water to your garden plants during a drought, and they will not thank you for it; they will wilt, because the mere 3½ per cent of salts render the 96½ per cent of water, in which the salts are dissolved, wholly unavailable to land plants.

Putting water beyond the reach of the processes of microbial chemistry is the basis of all the traditional, and some modern, modes of preserving food. These include drying, salting, and letting the food first go bad in order to keep it good.[1] Whether salt, sugar, or acid is added, or whether acid is generated by fermentation,[1] or whether actual water is removed from the sphere of action by drying the food or turning part of its water into ice, the net physico-chemical effect and metabolic consequence is the same (for equal reduction of free energy).

Examples of operation of the second method are: dried fruits, salt fish, sweetened condensed milk, ham, salted hay, pickles, honey, jams. In many preserved foods two or more factors operate to put water beyond the effective reach of

1. Making food bad in order to keep it good is a very ancient method of temporary preservation. Some such foods (e.g., cheese) will keep for long periods; other examples are liquid soured milks, and silage and sauerkraut. If merely fermented, foods are liable to go bad very quickly after exposure to air, since they retain a high percentage of total and available water. The usual fermentation-products, besides gas, are acids or alcohol; microbes are not able to produce enough of either to reduce the free energy sufficiently to arrest any kind of growth in presence of air. Fermentation usually implies exclusion of air (oxygen); thus, any full discussion of food-preservation – as of growth in soil – would include consideration of the roles of oxygen and air as well as that of water. It is thought that to attempt that now would take us too far from the principal subject.

microbes. For example, jams are boiled, are acid, and contain added sugar.

It is common knowledge that the second method is effective in preserving many kinds of food, so long as water is not allowed to come in from the air (a damp cupboard?) or otherwise. There is the familiar caution about not taking a wet spoon to condensed milk; and experience tells that condensed milk or honey will normally 'keep' satisfactorily under far from aseptic conditions.

Such facts indicate that foods preserved by the second method do not go bad primarily because of moulds or other microbes. The microbes are naturally present anyhow, both in the food and in the air.

They will, however, remain undetectable as spoilers (except by appropriate microbiological investigation) unless water comes in to reduce the concentration of dissolved substance – possibly only locally and inadvertently. A proportion of water then becomes available for growth, and the microbes declare themselves.

As an instance of the importance of *conditions* in deciding the pullulation of microbes or other organisms, the subject of food-preservation may have more general appeal than a dissertation on, say, the microbiology of soil as expounded in Chapter Seven.

When honey or jam has been allowed to get damp, its moulding or spoilage does offer a very good ecological text against the germ theory of *causation* of disease.

Spoilage of food preserved by methods (1) and (2) has for its immediate cause the presence of biologically free water. When method (2) has been properly used, the spoilage is not primarily due to microbes present in the food or the air. Without water at their disposal, microbes in food may remain alive but will certainly be passive.

The Dutch microbiologist L. Baas Becking said: 'Everything is everywhere; the environment selects.' That is natural history in six words.

CHAPTER SIX

Hills and the Sea

One evening he asked the miller where the river went.

'It goes down the valley,' answered he, 'and turns a power of mills ... and then it goes out into the lowlands, and waters the corn country ... until at last it falls into the sea.' – R. L. Stevenson, *Will o' the Mill*

THIS quotation represents what I may call a miller's view of rivers. It is strictly limited to the water, to the number of mills the water turns and can be supposed to provide work for. Mills need not be corn mills. For a while the Industrial Revolution depended much more upon water than upon coal. The famous Carron ironworks in Stirlingshire, which by about 1800 was the largest factory in Europe – and consequently in the world – was sited where it is because of the suitability of the River Carron for power, and the abundance of wood nearby for fuel.

The miller's view is practical, if also poetic, and is partly erroneous, being seriously at fault in one respect. It appears to be uncritically shared by many besides millers; including most writers who comment on the geography and natural resources of countries. The poet's view is no less molendinary in being limited to the water and what can be got out of it.

Inspiration may occasionally come from the fishes, in the way which delighted the young Lewis Carroll when he quoted from 'a German book'.[1] Fishes have 'ordinarily angles at them, by which they can be fanged and heaved out of the water'.

More often, the poet delights, Milton-like, in the names; as Robert Burns pointed out in deploring the neglect of

1. I am indebted to *The White Knight*, by Alexander L. Taylor (Edinburgh: Oliver and Boyd, 1953) for this citation.

poets to sing of the lesser-known rivers such as the Ayr and Lugar:

> Th' Illissus, Tiber, Thames, an' Seine
> Glide sweet in monie a tunefu' line;[1]

in the poem which also says:

> The Muse, nae Poet ever fand her,
> Till by himsel' he learn'd to wander,
> Adown some trotting burn's meander.

This poem was addressed to a local worthy of the district of Ayrshire known as Kyle or Coila, and may have been written in a wooded spot overlooking the River Ayr at Auchincruive which Burns used to frequent. This wood adjoins the grounds of my College, and, like those grounds, formed part of the policies of the estate of Richard Oswald, whose Adam house (built 1767) is now a hostel of the College. This Oswald was the only British negotiator of the draft terms of the treaty of peace with the young United States. Traditionally, the wood was a hiding-place for Wallace; and a monument in it is probably unique in commemorating two national heroes separated by five centuries. Not far away is the reputed burial-place of Old King Cole, a Pictish chief who ruled Coila; so this poem and the River Ayr together have called up a rich collection of history including the earliest official linkage of Britain with the United States as a political entity.

'Adown some trotting burn's meander . . .' Whenever the Muse who inhabits rivers and burns and lochs and lagoons is kind to poets or anglers or historians, she seems to be

1. Compare Giosue Carducci, *Alle fonti del Clitumno*:

> 'Salve, o serena del' Ilisso in riva,
> o intera e dritta ai lidi almi del Tebro
> anima umana!'

The name of this little river is seldom to be found in gazetteers or on maps. The *Oxford Companion to Classical Literature* (Oxford, 1946) mentions the Ilissus as descending from Mount Hymettus towards the south-east and south of Athens.

Burns's spelling is taken from 'Robert Burns', *ed.* H. W. Meikle and W. Beattie (*Penguin Poets*, 1946).

hardly more than semi-competent towards her literary auditors – or shall we say, executors? She reveals little except the water. Foaming or placid, the water is not just water on its way to the sea or other sink in which water can lose itself: it is a solution of air and of soil: a carrier of both air and soil. It is much more than a downward progression of water. It is distinguished by much more than a simple difference of level.

Rivers and burns fulfil at least three functions worth attention by people who are not anglers. Two of them relate to the river as carrier of air and soil; these two can be dealt with briefly and illustratively before passing on to the third and greatest: namely, the effect of water, and of rivers as the by-product of water, in determining the types and style of biological successions on land through the degree of aeration induced in the soil.

Let us take first an instance of a biological succession in the water of a river. The rapid or slow progress of the water will be accompanied by more or less solution of oxygen from the air, and by greater or less consumption of oxygen by the fish and miscellaneous microbial inhabitants of the water and the mud of its bed and banks; also by an output of oxygen by water-plants. With all these factors operating along any given change of level, it will be clear that the gaseous and biological equilibria of the river will be infinitely complicated. The natural tendency will be towards oxidation of effete organic matter coming from the life of the river itself and from the fall of leaves and other sources of organic matter on land; and towards oxygenation of the water if the river is able to pick up more oxygen than is needed for the decompositional work of its microbes – which are very similar to those of soil, and have the same ultimate aims of restoring to circulation the constituents of effete organic matter.

Now suppose the river to receive an abnormal load of organic matter, say of sewage from a town or of sugary wastes from a factory or of milk-washings from a creamery. The principal effort of the river's micro-organisms will be directed to oxidizing this carbonaceous matter. In the process they will multiply enormously in numbers, locking up for a period a good deal of dissolved nitrogenous compounds in the proteins of their own tissues; but they will also respire, taking up as much oxygen as possible. The balance between oxygen available and oxygen consumed may come at any point, but if the amount of added 'foreign' organic matter is very large the result can only be an almost complete exhaustion of dissolved oxygen. This may mean a supersession of the oxidizing microbes in favour of the species which do not require oxygen gas (anaerobes, they are called). Thus microbial life does not stop, but is likely to be profoundly altered – and to be accompanied by bad smells and other signs of predominance of anaerobes and of a lack of dissolved oxygen. In the worst outcome, the fish will be suffocated; and so may the water-plants be; with the result that more dead matter is added to the burden of the stream.

However, the water is flowing downwards, and is still picking up oxygen from the air. What will be the net result of all these changes in level, in amount and kind of organic matter more or less decomposed, in oxygenation, and so on? One cannot tell; 'it all depends'.

One thing is certain: there will be a biological succession – much more marked in polluted rivers than in streams in a natural condition – but always a rapidly-changing succession: varying not only downwards along the direction of flow of the river, but also from side to side (since the velocity and turbulence of the water will not be uniform), and from the surface downwards. The water microbes and the fish and plants will, as far as they are able, take part in a natural purification of the water. The fish and plants may at times and for a space be outed, but the microbes never: they will adapt themselves to any and every condition in their eternal job of decomposition and re-solution in a stream, as ineluctably as they do in soil. And, precisely as in soil, the microbial equilibria will depend upon, and correspond exactly to, the amounts of energy-material which are available to the microbes, together with the amount of air-oxygen utilizable for the principal elements of dead organic matter to be oxidized to carbon dioxide and water.

Broadly, the living inhabitants of a stream are in balance during their natural struggle for life; but it is a strange thought that as one walks downstream from a point whereat pollution is received by a river, the invisible flora and fauna change in relative and absolute numbers with every step one takes.

All soil and its products are on the way to the sea. All soil is on its way to the sea, and mostly by river. The Venetians have known this for many hundreds of years, at least inasmuch as it affected them: practically since their Republic was founded on the Rialto (*Rivo Alto*) in the middle of the lagoon, more than a thousand years ago.

Historians are not slow to point out the declension of ports like Sandwich and Rye, from which the sea has retreated owing to an advance of the land. Yet it is rare to find an account of the power and glory of Venice at her height of empire, or her present not inconsiderable importance as tourist centre and port, which treats of these otherwise than as an affair and consequence of enterprise with ships and the sea, and of naval and military and commercial competence and fortune.

The very security of Venice as a colony and city; her existence as a maritime power; all rested on the ability of Venice to prevent her lagoon from being filled up with silt brought into the lagoon from the nearer Alps and hills by the rivers Bacchiglione, Brenta, Sile, Piave, Zero, and others. Still other rivers, of which the Adige and Po are the most notable, threatened to 'wrap up' the lagoon from south and north with silt from more distant origins in the crumbling mountains. Silt from the land would make the lagoon unnavigable; Venice would not be a sea power if the lagoon filled up; and the Venetians would be vulnerable and poor if they did nothing, or less than enough, about it.

The fleets of Adria and Classe have vanished, because the people of those towns did not learn their lesson about rivers as well as did the Venetians. The fleets of Classe and of

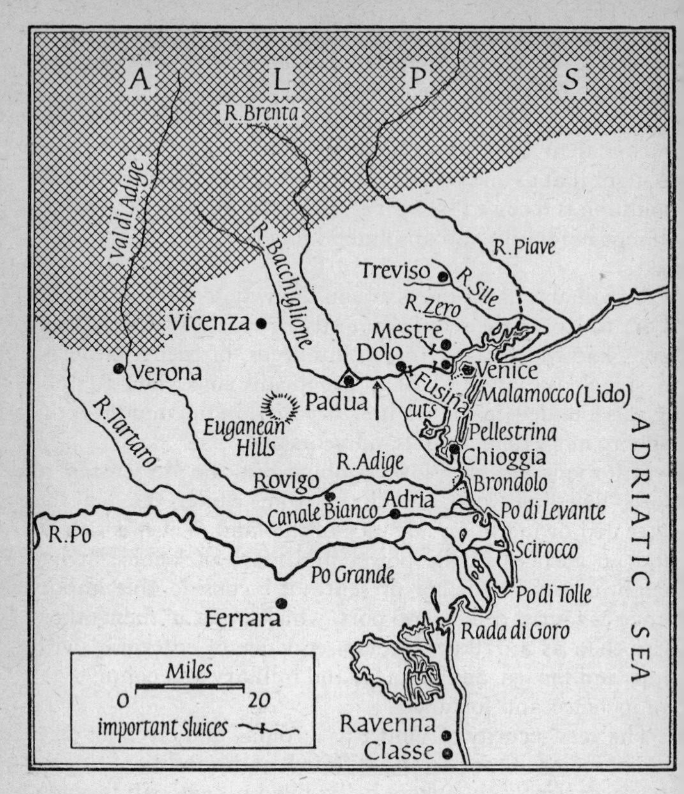

Fig. 1. Sketch-map to show the principal area of river-training and mud-control by the Venetian Republic and its successors.

For the River Piave, which was given a mouth to the east outside the lagoon of Venice, the former course into the lagoon is shown dotted. The Brenta may be regarded as in part the water-course so named (arrow) between Padua and Fusina; but most of its water is now taken by artificial channels towards the former lagoon of Bròndolo – now filled up by mud – and other outlets to the south-east. Two of these cuts are shown.

The Rivers Brenta and Bacchiglione form a network of waterways in and around Padua. Like most of the other lowland rivers the Sile once had a different course from that shown on this or other modern maps. By about 1840 the Adige had seven mouths; it now has only one.

Dolo is one of the main regulating-points of the Brenta; an eighteenth-century painting showing its sluice is in the Ashmolean Museum at Oxford. Fusina is a terminus of the electric tramway alongside the Brenta from Padua which connects with the ferry giving a fascinating approach to Venice itself. [*The Merchant of Venice*, Act III, Sc. 4, ll. 53–4.]

Venice long ago sailed together; but one vanished, the other grew and spread out over the known seas.

Venice knew, or divined, that to remain Queen of the Seas she must become effective Mistress of the Mud. So for centuries Venice has toiled with engineering works on the hinterland as far west as Padua and beyond, and as far south as the southernmost mouths of the Po, and to the north and east as well; digging, diverting, and draining. Some of these immense and long-continued operations were applied to food-production, others to inland navigation no longer important; but these purposes were secondary to the preservation of the lagoon as a harbour and a moat.[1]

Do you know of any description or analysis of the history and glories of Venice which has not concentrated on her overseas trade, her crowded palaces and churches, the romantic story of her islands and paintings: without a hint of enquiry about why the Venetian islands stayed as islands in a region where water was fast becoming land, except in Venice? It was not by accident that the lagoon of Venice has lasted as a lagoon until our time, while other lagoons nearby have filled up and left hardly a name.

That historic activity of Venetians which made Venice famous could not have persisted without equally protracted and heroic attention to the mud pouring in at the back doors. The front door – leading to Constantinople and the East – is celebrated; but the back door was no less important. So Venice mastered, and regulated, the rivers in and near her lagoon, as far as the Po, which was partially diverted: lest the very basis of Venice's life and prosperity, of her existence even, should be filled up with soil material and become dry land: as happened to Classe, a seaport now

1. I have been fortunate in getting information about this fascinating subject of water- and mud-control by Venice; through the kind offices of Professor Viscardo Montanari, Ing. Luigi Vollo supplied me with details which would otherwise have been very difficult to obtain. Ing. Vollo is attached to the ancient Magistrato alle Acque of Venice. He has published a résumé of some of his historical knowledge, under the title 'L'idraulica veneta nel Rinascimento', in *Tecnica Italiana* (*Riv. d'ingegn. sci.*), Trieste, 1948, No. 4, 3.

represented by an ancient church on what must have been a little elevation (a *rivo alto*) now level with surrounding fields, and a mosaic in Ravenna showing towers and ships once possessed by Classe. Adria is fifteen miles from the sea to which she gave its name. [Fig. 1, p. 74.]

Compared with the little-known efforts of the North Italians in their struggles against water on its way to the sea and in more positive aspects of utilization of water, the much-publicized activities of the Hollanders seem dull. After all, the Dutch have been mostly concerned with getting rid of water, and with keeping it out. The northern Italians have practically every manifestation of water-control and utilization; from torrents to salt-marsh. There are geological erosion and soil erosion; fresh water and salt; snow and salinity; winter rain and summer drought; and almost every form of irrigation is to be found in north Italy, as well as reclamation of most kinds.

To anyone interested in the roles of water in agriculture, food-production, civilization, and history, northern Italy offers a compendium of examples. Some of these are the finest: thus Venice as the prime example of the dependence of sea-power upon the land and its management. A special kind of irrigation is mentioned in Chapter Sixteen; and still another Italian mode of control of water is the immemorial system of agriculture based on *baulatura*, which has never (as far as I know) been described in English. Such neglect is hard to understand.

Nor can I understand why so many people speak of the land being watered by rivers. Rivers do not water land except by flooding; though, as we have just seen, rivers can turn water into land. What is called a 'well-watered land' may be copiously rained upon; and to that extent the occurrence of many small rivers in a region indicates a local surplus of water: an excess of precipitation over evaporation. Conversely, poverty or absence of streams must indicate an excess of evaporation over precipitation. One big river, or river-system, like the Nile and the Indus, often implies no more than the existence of heavy rain or snow somewhere else.

Except by flooding, rivers, if untouched by man's scooping or pumping or other diversion of the water to a level above that natural to it at a given place, do not supply water to land: they supply air. Rivers do not water soil; but they do aerate it. Normally, that is, and so long as they are flowing (with the water in motion). The river's sink is usually the ocean, but may be an inland sea or swamp. Whatever the sink, any soil above it will have been aerated by the downward and riverward flow of water through the soil. This water-aerated soil may be naturally above the natural level of the water; or the soil may be raised, at least in part, above the level of the water (as in *baulatura*); or the level of the water may be lowered by drains and ditches; or the water may be raised above its natural level and allowed to return to it, as in most forms of irrigation by flooding and sprinkling; and there is subterranean irrigation – reversed drainage, so to speak – in which water is periodically allowed to rise into soil from underneath.

Is irrigation, then, simply a giving of water? Or has it more complex functions, in some of which the supply of moisture is no more than incidental, so that the water has essentially a mechanical function? This will be discussed in Chapter Sixteen.

All agriculture depends upon the management of water, according to whether the water for plant growth is inadequate or excessive. If water is inadequate for growth of at least one kind of crop, there can be no agriculture. Where water is, on the whole, adequate for only one kind of crop, as in the semi-arid wheat lands of North America and Australia, agriculture tends to take on a character of mining the land, sometimes with no return (as in North America). Where water is inadequate or barely adequate for growth of one crop or a rotation of crops, the air-supply in the soil will be good. Where water is slightly or very much in excess, there will be periods during which the soil will have too much water, and consequently not enough air. It will therefore be necessary to undertake drainage artificially – nominally to get rid of the excess water, but mainly to allow a larger proportion of air to enter the soil; or else the consequences of reduced productivity of the soil must be accepted.

When a soil is drained artificially, whither must the drainage go if not either to a river or direct to the sea? And this drainage-water, has it not let in air as it went out of the soil, as it began the first stage of its journey to the sea? And will not natural drainage behave similarly, if the water is free to run down to the sea?

The case for aeration of land by rivers rests.

The case of 'watering' of land by rivers is a little more complicated to argue. If a flood occurs through a spate of water from rain or snow at a distance from a given area, that area may be flooded visibly through the rising of the water above the level of the soil, or it may be 'flooded' or moistened invisibly by underground seepage of water to a relatively small distance on either side of the river, if we suppose the banks to be permeable (not, say, massive rock). If little or no rain has fallen in the area under consideration, such flooding or underground seepage will have truly watered the soil. But flooding is undesirable, and if it occurs often is incompatible with general farming. Lands apt to be flooded tend to be reserved for grazing; and thus imply the existence of arable land from which the animals can be fed in supplementation of the grass on the low-lying fields.

The level reached by underground water is called the water table. Sometimes the water table rises above the soil, so that the soil is visibly flooded; but we will restrict consideration to a water table actually beneath the soil. If a river has permeable banks, the water table will have broadly one of two forms. Before showing those, it is necessary to explain that a water table is hardly ever level: it usually *slopes* perceptibly either downwards or upwards. This sounds queer, because (curvature of the earth excepted) the surface of a lake is always perfectly level. Acceptance of the idea of a sloping water table may be eased by recalling that a river or stream or a ditch with water in it is hardly more than the visible equilibrium-level of the water table in the surrounding soil (if the land is 'well watered'); and the slope of the watercourse is downward.

The idea of a water table sloping upwards is more confusing. One cannot imagine water heaping itself up to an appreciable height! The 'upward' slope is simply a matter of terminology, or of the way of looking at things. Water draining to a river is always going downwards, so if one looks along the direction of the flow the slope of the water table is always downwards, and common sense is not confuted. But if one looks or works or measures away from the river towards the slopes of the land on either side, then any water table due to drainage from those slopes rises as one gets further from the river, so that it may be said to slope upwards. [Fig. 3.]

The two principal conditions of the water table in relation to a watercourse are shown in the diagrams on p. 80.

Only in the condition exemplified by Fig. 2 can a river be said to water the land, and obviously only to a limited distance; the water will never be above the level of the river if that is in steady flow, and will usually be below it (unless the river's level has suddenly fallen). Where the land adjacent to a river is only a few inches or a foot or two above the level of the river surface (*i.e.*, where the water table is high) the land will be fit only for grazing even in the unlikely event of its not being liable to flooding. The high water table will not allow most crops to send their roots down into aerated soil for a sufficient distance, so they cannot be grown – other considerations apart. With a high water table, air in the soil is the limiting factor of growth. Complications arising from tidal variations in rivers need little more than mention here.

Use of underground water to keep soil both moist and aerated is well shown by the Dutch practice of adjusting the water table (by control of the level of water in ditches) to various heights to suit different crops – including tulips and other bulb joys – on various types of soil.

Raising the level of the soil above (tidal) water is the basis of the operation known in English as warping (French, *le*

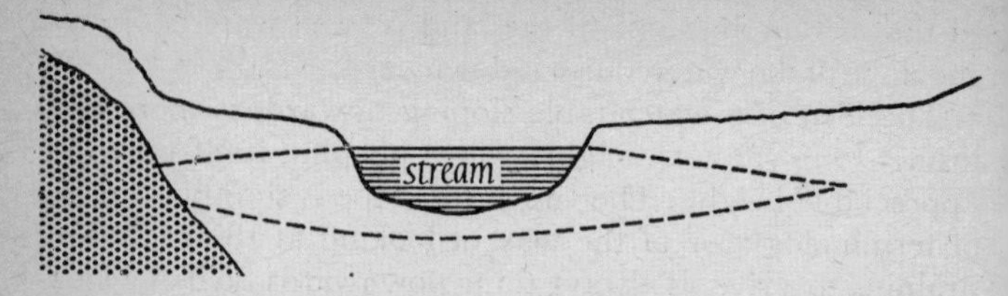

Fig. 2. Distribution of water underground *from* a river into permeable soil on both sides, no recent rain having fallen in the area. The water table slopes downwards from the river.

colmatage; Italian, *la colmata*). This is done by constructing a channel from the river to enable the high tides to come in over an embanked area, upon which the quiescent turbid water deposits its silt, and flows back quite clear as the tide falls. The operation is not as fool-proof as this simple description makes it appear; the highest tides may not be admitted, and consequently there must be sluices and some watchfulness. Twice-daily inflow of the majority of tides deposits silt at a surprisingly rapid rate; a couple of feet of soil may be built up in a couple of years or so. Around the

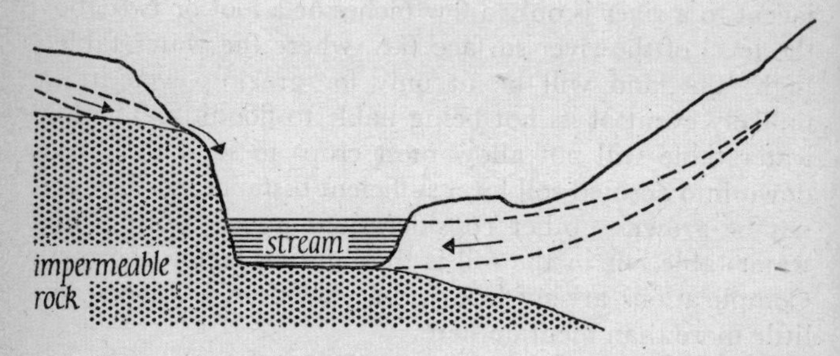

Fig. 3. *On the right:* Distribution of water percolating through soil towards a river, shortly after heavy rain has ceased. For simplicity, the river is given a rocky (impermeable) bed, so that the effect shown in Fig. 2 is omitted. The water table slopes upwards from the river.

On the left: A 'perched' water table, *i.e.*, one well above the general level. It may give rise to a spring (arrow).

Humber, for example, much of the arable farming is done on warped land; but unfortunately there seems to be no more land in England suitable for warping. A few years ago I saw what was said to be the last possible field being warped in Yorkshire.

Baulatura is a much more laborious operation for the same purpose of raising soil above the water table, and thus aerating it to fit it for growth of arable crops – especially maize, lucerne, wheat, and sugar-beet. It is practised over thousands of square miles in the plains of Lombardy, Venetia, and Emilia; the towns of Vicenza, Padua, Rovigo, and Ferrara may be thought of as centres. The word *baula* may be translated as hog's back or hummock; and *baulatura* consists of laying out the fields in comparatively narrow ridges running the whole length of the field. The ridges have a height of about a metre; the hollows between serve as surface drains, and are connected to main drainage channels. Mulberries and other trees, often with vines trained up them, are grown along the margins of the fields.

The principle of *baulatura* is much the same as that of the British ridge-and-furrow which used to be common on the less permeable 'heavy clays'. The British ridges were ploughed out, *i.e.*, formed by ploughing in a special fashion so as to throw the soil away from the 'furrows' which were to serve as surface drains, and to amass it into ridges or riggs. The word 'furrow' here must be understood in a more specialized sense than that of the single furrow cut by a plough in the ordinary mode of ploughing in preparation for a crop. When the ridges had been made, they were ploughed and cultivated in the normal manner before the corn or other crop was sown. Most ridge-and-furrow fields were laid out long ago, in the days when animals provided the only motive power; and the necessities of turning the teams at the ends of the ridges led to the ridges taking the shape of a long S. Such more or less wavy bands can still be seen; if they are in a grass field, they imply that the field was once sown to arable crops – perhaps a century or more ago.

The Italian *baula* is straight, having been made by spadework, possibly many centuries ago. The transformation of a large area of originally flat (alluvial) land into the immense number of baulks or ridges it now carries has been described as the greatest feat of earth-moving ever performed by man. Whether that is literally true I do not know; but it is almost certainly the greatest single feat of hard manual labour in Europe, if not in the world.

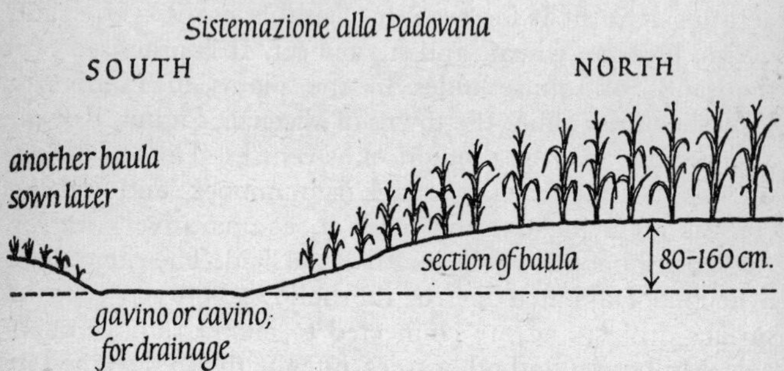

Fig. 4. Diagrammatic representation of maize growing in early summer on a *baula*, seen from the side. Only a single row of maize is shown.

Since the north Italian climatic régime consists of rain in winter and drought in summer, *baulatura* represents a compromise, and is sometimes complemented by summer irrigation. The prime purpose of *baulatura* is to lift the crops above the water table of winter and spring. The main mass of earth in the ridge carries enough water for the growth of quick-maturing shallow-rooted crops like maize; but, near the ends and bottoms of the ridges, maize may not get enough water when the drainage-channels have dried out. Consequently, growth on the ridge is often uneven, the maize at the bottom being stunted, the plants near the top being tall. The tops of the plants exaggerate the slope of the ridge.

An observer who knows nothing about the peculiarities of Italian agriculture can recognize the system of *baulatura* from a train passing at speed through a region where it is prevalent, especially if the ridges run parallel to the railway (they tend, in fact, to run east and west or north and south), Maize is perhaps the best indicator. It looks very much like Fig. 4, if allowance is made for the whole length of the ridge not being shown in the figure.

We thus have some striking exemplars of the importance of aeration for crop production. The Wharfe or the Forth or the Po will not have watered the land except to the extent that it has flooded it in winter. Where there is not enough water, man may be at pains to provide it or to store it up; but far more hard work has gone into getting rid of water or otherwise letting air into soil, than has been put into irrigation for the purposes of letting water in.

It is possible that the story of King Canute telling the waves to go back has an origin quite different from the fable that he gave his courtiers a lesson about the impotence of kings. There is reason to suppose that he initiated a scheme of reclaiming land from the sea near Bosham in Sussex by construction of a sea-wall. Perhaps the wall broke; we don't know. If it did, he had failed to keep the waves back; but, whatever the outcome may have been, it is good to remember that he had tried to let in air as well as to keep water out. If Canute ever tried to reclaim land from the sea, the version of that effort which is current to-day has got a perverse twist, since it mentions only the water. So do the millers and the poets and those who talk about land being 'well watered' by rivers. In general, rivers drain land; they don't add water: they take it away; and, in doing so, let in air.

Matters of high consequence arise from these facts.

CHAPTER SEVEN

The Grand Principle of Soil Microbiology (and of Soil Classification)[1]

'And what is the sea?' asked Will.

'The sea!' cried the miller. 'Lord help us all, it is the greatest thing God made. That is where all the water in the world runs down into a great salt lake.' – R. L. Stevenson, *Will o' the Mill*

WE continue with another only partly appropriate quotation. There is a better one in the reference by Gerard Manley Hopkins to the 'wild and wet' in the poem about Inversnaid, where the hills come down steeply to Loch Lomond and the shore is mostly a sharp discontinuity between rock and soil and water: between solid and liquid – though not, in this instance, between fresh and salt.

* * *

IF we descend in imagination from the highest summits to the sea, we pass from a region where there is no liquid water through regions where liquid water may exist as a constituent of soil or as swamps, lakes and rivers, down to the sea. Like lakes and rivers, the sea is a medium for life only in so far as it is liquid.

The transition from soil to a mass of water such as river, lake, or sea is at first sight striking, and is apparently discontinuous (the special case of swamps will be dealt with later). The appearance of discontinuity of biological conditions vanishes if a few points are borne in mind. These are principally:

(i) Soil is a cellular rather than a granular mass: the spaces

1. And, as an afterthought, it is the substratum of every other biological thing that depends upon soil and microbes and plants. Us, for instance.

of soil are biotically much more important than the particulate solid matter;

(ii) Hence in contrasting soil with water as media for life it is fallacious to regard the one as solid and the other as liquid.

(iii) There is no primary difference between general biotic conditions in the soil and the ocean; both are aquatic media requiring good aeration for development of higher life and the normal prosecution of the life-cycle.

It will be shown that in all natural aquatic media, aeration depends upon the movement of water; and, to a first approximation, the intensity of oxygenation varies directly with the extent of motion of water. For soils, the totality of this motion is not simply flow, but is a more complex and interrupted coming and going.

Point (i) is after Dr H. L. Penman, of Rothamsted Experimental Station. Rothamsted work on the physics of water in soil has been especially fruitful. Point (ii) is but one of the consequences streaming from the newer view of soil as a system of spaces rather than of particles.

Point (iii) needs elaboration.

The chief differences between soil and ocean as media for life are (*a*) no biological fixation of atmospheric nitrogen has been shown to exist in the open sea; (*b*) the seas (and other masses of water) though layered, are structureless in an ordinary sense, whereas soil possesses several kinds of structure.

(*a*) The sea gains combined nitrogen from the washings of land, and there is doubtfully a further small gain from *Azotobacter*-like nitrogen-fixing bacteria living in loose association with sea-weeds. The open ocean, however, seems to have no other source of combined nitrogen than atmospheric precipitation. (In some lakes bacterial nitrogen-fixation may be important; and nitrogen compounds leached from the soil must be relatively more important in lakes than in littoral zones of the sea. In a short discussion, however, the special details of conditions for life in lakes and most other nominally aquatic media which are in contact with soil may be left on one side.)

(*b*) The contrast between structural soil and structureless sea, lakes, rivers, and swamps is not crucial. What is important is the degree or extent of *interchange* and renewal of water and gases. In the ocean where the amount of water is virtually infinite, exchange of water with the outside environment is quantitatively negligible as a biotic condition; but motion and circulation ensure that there shall be no stagnation. In soil, seas, and lakes, downfall of rain has a secondary but important effect in directly facilitating exchange of gases. In the ocean, waves are the main mode of effecting exchange and renewal of gases. In soil the vital interchange and renewal of gases is also operated by water.

The importance of *interchange* of gases between the biotic medium and the atmosphere is well brought out by consideration of swamps. In swamps, where there is virtually no interchange, a biological association is set up which is quite different from that of any other medium (excepting such special examples as limans[1] and pond bottoms and the like). In true swamps[2] and limans no higher forms of life can exist: because all high forms of life depend upon gaining oxygen from the air. Where there is abundance of water, but no substantial interchange of water or of gases, only anaerobic bacteria can flourish.

Where there is abundance of air, but no available water, higher life cannot, again, exist. Such a habitat is dry soil;

1. A liman is a sort of marine backwater, along the coasts of the Black Sea especially.

2. Wet-rice or *padi* fields, sometimes called rice swamps, are not true swamps, since the water is not stagnant. It is kept moving, and usually in two directions – over the soil and away into ditches at the sides of the fields, and also percolating through the soil. Algae evolving photosynthetic oxygen may provide accessory means of aeration. With some diffidence, I suggest that the principal function of the water on wet-rice fields is to act as a carrier of oxygen rather than of moisture; or at least to supply oxygen both directly by carrying oxygen in solution and indirectly through encouraging the growth of algae.

and the only forms of life that can exist in dry soil are certain microbes (and seeds) specially adapted to resist drought, though possibly unable to flourish in extreme conditions of dryness.

The operative factor for life in presence of abundance of air is not the percentage of water present, but the availability of the water. The percentage or content of water in soil (also in the highly salt habitats, which lie outside this discussion) has no biotic meaning, but is a local and historical accident depending upon a variety of trivial features such as the length of time since precipitation last occurred, the amount and kind of precipitation, factors governing the loss of water, and so on.

The availability of water in soil has two facets. One has already been mentioned: the facility of gaining water from outside the soil, and of losing water to the external environment. This comes under the heading of interchange, and is of the greatest ecological importance. The other facet is what plant physiologists and soil physicists usually mean by availability of water, and is most compactly expressed by the pF concept. A distinction must be made between soil water available for plant growth, and soil water which is not available for plant growth. This availability of soil water for plant growth is diminished – for similar percentages of total soil water – by acidity, by presence of salts, by freezing (which removes water from the liquid state), and by adsorptive effects of soil colloids. pF is a function of suction force upon which plant growth depends in relation to soil water. Though the pF concept has perfectly general validity within its sphere, and is by no means a static concept, it is less dynamic than the notion of continual renewal of water in soil that is brought about through circulation of water in the evaporation-precipitation cycle.

Soil water ceases to be available to plants when the pF is about 4·2–4·3, corresponding to a suction force of some fifteen atmospheres. Any water held by soil at a pF greater than about 4·2 is not available for plant growth. No soil is perfectly dry at pF 4·2, and many soils are moist or even wet – as judged by 'feel' and appearance, or percentage data – at that pF. (Some peats may contain 40–50 per cent of water at pF 4·2: that is to say, a peat may contain water amounting to about a third of its gross weight, all that water being unavailable to support plant growth.)

It follows that in soils at any given moment and at any given depth in the soil much – perhaps most – of the water in it is physically unavailable for growth of rooted plants, for microbial growth, or for sustaining animal life or other consequential growth. Nevertheless, soil is essentially an aquatic medium.

It is also true that in soils most of the soil air present at a given moment will, if not renewed, be unavailable for the purposes of aerobic organisms macro- and micro-. Though there are some special points in connexion with use of soil air (including the point that anaerobic micro-organisms are not necessarily independent of air, but only of atmospheric oxygen gas) there is with soil air no analogue for pF of soil water; there is, however, an analogy, as far as respiration is concerned, with aeration and ventilation of rooms for human habitation. Soil air, like room air, requires to be frequently changed if the purposes of the highest organisms are to be served.

If interchange of soil air is imperfect, that oxidation of effete once-living organic matter upon which the cycle of life depends will be impeded, or, in the extreme case, stopped. We can conceive of all states from complete aeration (wherein the oxygen percentage will approximate to 21) to complete anaerobiosis or lack of gaseous oxygen. The latter state is equivalent to stagnation. The former state may be reached in soil, ocean, lake, and river – though the essential interchange of gases is wrought by very different mechanisms. In soil there is the further necessity of interchange and renewal of water from the outside.

Briefly, to obtain optimal growth in and on soil it is necessary that air and water shall both be *in passage through* the soil. Assuming that nutrient conditions and other physical conditions such as acidity and salinity are also good, satisfaction of this condition assures optimal activity of the whole corpus of soil microbes and also of plant growth.

States of soil varying from complete aeration to stagnation may vary in time and in place, or both together. The soil has a complex structure: not only does its structure vary somewhat grossly in depth, below any one point or line on the surface, but there are finer variations in small volumes owing to the existence of crumbs, or aggregates of particles of varying size, which are to some extent discontinuous. Thus there can be variations in composition, and most notably in such physical features as absorption-capacity of the inorganic and organic colloids, and hence in pF, permeability, and so forth: and such variations may occur in strata ('horizons') several centimetres or feet thick, and also very locally across the diameter of a crumb less than a millimetre in radius. Apart from gross changes such as result from the formation of visible cracks, there can be wide variations of wetness (total and/or available water) and in degree of aeration and aerobicity, from place to place in a soil. Moreover, the régime of water and air is constantly changing in a soil with time.

It is necessary to bear in mind the existence of such local and temporary variations, but a consciousness of them is not necessary for the most fundamental view of the relation of soil water to life.

Pursuing the idea with which this Chapter opened, and descending a mountain from a point just below the snow-line (ignoring variations in that, and for the purpose of considering only the effects of liquid water – however that water is brought), it will be found that only exceptionally a continuous descent is possible. Sooner or later a downhill slope changes into one which man, in walking towards the sea, will regard as an uphill slope. Yet, for water, slopes are all downhill, whether we look on them as facing the sea or leading temporarily away from it. Almost the only extensive and continuous slope that can be found in nature is the narrow geographical feature called a river: though it may be a *wadi*, or, exceptionally, a dry valley. Slopes other than rivers or river-valleys, or, in other words, slopes which may with sufficient accuracy be termed 'at right angles to water-courses' are uncommon over distances of more than a few hundred metres. The justice of this may be seen from considering vertical 'slopes', *i.e.*, cliffs. The highest cliffs in Britain are in the Orkneys, and are about 600 feet high; the highest cliffs in the world are said to be on the eastern face of Formosa, and are a few thousand feet high. There may be volcanic screes a few thousand feet long; but sufficient has been said to indicate how far one may have to go to find a long continuous downhill progression of rock covered with soil.

Thus, for the argument about the ecological consequence of water, it is hardly necessary to consider a slope of which the total rise is more than a couple of hundred feet. Much larger continuous rises, however seldom they exist, must bring in the effect of altitude upon climate. The catena concept of G. Milne does include such altitudinal changes of climate, soil, and vegetation, but is not based upon continuity of slope. The present argument is restricted to the effect of a continuous slope of moderate texture and appreciable gradient, having negligible change of climate from top to bottom, and assumed to possess substantially the same soil – more accurately, the same soil parent material – throughout. Such abrupt transitions as, for example, from soil formed from sedentary rock or glacial drift near the top, and alluvium at the bottom, are excluded from the general argument; where such transitions exist, soil of each origin must be separately viewed.

The effect of liquid water coming (most commonly as rain) upon the top or any part of such a slope will initially depend very much upon the previous history of the soil. (It is assumed that the soil is virtually uniform, but the gradient need not be uniform so long as it is all downhill and without lateral transgressions able to form an incipient stream or burn of much consequence. After this caveat, such a slope will be called a 'uniform slope'.) The most general case of wetting will be by rain, which can also be assumed to have uniform distribution over the slope. If the top of the slope is wetted beyond its capacity to absorb water for any reason (though it is not necessary to assume always that the soil must be wetted to saturation) the unabsorbed water will run downwards. There are two components of the downward motion of water through or over soil on a slope: the vertical, which must be through the soil, and a horizontal one, which may be either over or through the soil, or both. Thus, even if run-off[1] is ignored, drainage can take place both vertically and also down the slope. Vertical drainage will proceed from saturated soil until the descending water meets an unsaturated but permeable layer, so as to increase the depth of soil at its field-capacity – in which case the water will be absorbed and retained; or else the water will meet an impervious layer – in which case it will join in the flow down the slope. It will therefore be sufficient to consider flow down the slope.

After a cessation of rain heavy enough to cause the soil to drain, a possibly discontinuous soil-water equilibrium will be established. If the rain is not very heavy the soil at the top may soon be left only 'moist' while that near the bottom is saturated. If heavy rain falls again shortly after, a continuous, though temporary, soil-water equilibrium may be established over the whole slope. Infinite variations are possible; yet it will be seen from this simple example that the soil near the top may have enjoyed an interval – however brief – of renewal of both water and air, while the soil near the bottom has had its water renewed, but awaits for a period the renewal of air. On a slope the water of even saturated soil will be flowing, and therefore renewed; furthermore, since the rain brings dissolved air, and tends to drive air into the soil, saturated soil need not, from purely physical points of view, be anaerobic.

Biological reasoning indicates, however, that the oxygen

1. Passage of water over the soil, owing to such high intensity of potential wetting that the soil cannot absorb the water, either through the soil's being already saturated, or through imperfect permeation for other reasons.

brought by rain is inadequate for more than very brief satisfaction of the demands of the aerobic organisms; hence, even at a little distance below the surface, continuously wet soil tends to be anaerobic owing to microbiological activity pursued by aerobes and facultative anaerobes as far as is possible for them. This lack of free oxygen, together with the presence of moribund organic matter of the unsuccessful microbes and plant roots, causes reducing conditions to be set up; and these are often sufficiently pronounced to induce change of colour of iron compounds. The result may be the phenomenon known as *gleying*, which is detected (after digging a hole) by the varied mottlings and sometimes brilliantly-coloured patches of the soil profile.

Gleying is an extreme condition arising from poor drainage. It is mentioned here because either it or bog- or swamp-formation is the terminal phenomenon of slopes receiving more water than can be drained away sufficiently fast to permit substantially complete aeration of the whole depth of soil. Obviously, all stages of aeration can exist for varying periods of time at various parts of a uniform slope. In northern Italy, for instance, a soil can suffer cool-humid conditions during winter and a virtual drought for several other months. In absence of excessive rainfall combined with low evaporation such as conduces to formation of peat, the general tendency on a uniform slope of moderate texture will be from freely aerated at the top to possibly nearly complete anaerobicity at the bottom.

The picture thus presented is similar to the concept of *hydrologic sequences* introduced into soil-survey work by J. H. Ellis of Canada. It was successfully used in Britain first by the workers of the Macaulay Institute, and later was adopted as a basis for the soil survey of Scotland under the general superintendence of the Soil Survey Research Board with the collaboration of the Macaulay Institute: in fact, the same workers have been responsible for both the British developments.

The concept of a hydrologic sequence postulates that the vegetation and the soils of a slope will be conditioned by the drainage, which can be expected to vary broadly from good at the top to indifferent or poor at the bottom. Consequently, the natural vegetation will adjust itself to these varying conditions of drainage. Further, since the soil and natural vegetation are conjoined and are ecologically interdependent, there will be variation in the *soil* between top and bottom of a slope. It has been assumed for simplicity that the soil on a uniform slope was uniform; but that over-simplified idea – necessary only as a starting-point – is untenable; the natural drainage will affect the type of soil that will be formed from an initially uniform parent material, and hence a hydrologic sequence of soils is formed. This hydrologic sequence or succession of soils is recognizable from its variation of pedological characters from top to bottom of the slope: quite apart from appeal to the varying vegetation borne by the slope at different heights.[1]

The hydrologic-sequence concept does not seem likely to displace wholly the original Russian 'great soil groups' – defined mainly on a broad climatic basis – which were the foundation of soil morphology and soil morphogenetical studies. The original Russian classification of soil types was based upon accidents of Russian geography; and experience has shown that neither the individual soils nor the succession of Russian great soil groups are quite duplicated elsewhere. The Russian soil types and groups are therefore local rather than world-embracing; and the same is true of most soils. That the Russian soil groups each occupy large areas of Russia, and that each shades satisfyingly into its neighbours, is no doubt due to the prevalence of similar climatic conditions over those areas. The hydrologic-sequence concept offers a rational alternative to more or less invalid attempts at fitting Russian types of soils into small areas of varied

1. If the slope is nearly level or very steep, special features will be presented which do not detract from the general ecological argument, which is here presented only in outline. I am indebted to Mr E. Bruce Mitchell for telling me of an alluvial 'flat' near Dalry, Ayrshire, which consisted of a very slight slope of almost impermeable clay covered with peat. In this instance the sequence from imperfect to impeded drainage was vertical.

climates unlike those of Russia; and, to valid classification of soils on the lines of the 'great soil groups', the hydrologic-sequence concept is a useful supplement.

In a hydrologic sequence, the emphasis is on the drainage; that is to say, on the water-relations. It is equally correct to regard the soils of a slope from the point of view of their air-relations, or better, their relations towards aeration or oxygenation. The spaces of soil must be filled with either water or air, or both. The volume of the spaces of soil being equal to the volumes of air plus the volumes of water: drainage affects aeration.

On any slope, therefore, there will be a progression of degrees of aeration. The microbiological sequel will be a progression of microbial activities – each a resultant of the interactions of species in differing populations evoked as responses to the environmental conditions. The microbiological progression or sequence will not be uniform in space or time even on a uniform slope. This absence of strict sequence arises partly from factors which have been mentioned but have been expressly neglected hitherto in the reasoning. The prime point is the fact that microbes are highly sensitive, *as mixed populations*, to their ambient conditions.

A discussion of microbial ecology in soil must take account of the fact that the soil microbes are a very mixed population. They comprise not only many species but many different kinds of organisms, including such diverse forms as bacteria, protozoa, yeasts and other fungi, and algae: as well as insects and other invertebrate animals. All these are necessarily at any one time and place in or approaching equilibrium – though it would pass the wit of man to say for any given case what the equilibrium is. In soil it is the equilibrium, rather than the population in any ordinary sense, that is highly sensitive to conditions. Among bacteria, for instance, the equilibrium can shift very rapidly – possibly within minutes; but the equilibria between the bacteria of soil are also conditioned by, and react upon, the other microbiological equilibria.

Microbes, being very small, are individually, and as populations, sensitive to minute variations to which larger organisms such as the higher plants do not obviously respond. Microbes are therefore sensitive to environmental variations like those existing between the periphery and interior of soil crumbs; such variations are set up within seconds or minutes after a crumb has been wetted, and they progress while it is drying. Minute variations in, say, wetting and therefore aeration, are almost immediately reflected in the balance of microbial (especially bacterial) types. These alterations in turn set up secondary variations, of which examples are the rate of disappearance of carbohydrate, the type of acid or organic colloid formed, the chemical status of iron, and the rate of production of ammonia. All these and other chemical transformations are performed simultaneously but at varying intensities by different species and under different conditions of aeration; and the chemical transformations lead to physical changes of which variations in acidity and ion-exchange are examples.

About such intimate changes evoked by small variations in conditions it is impossible to do more than generalize. But, because the equilibrium-principle governs microbial life in the soil, the discipline of soil microbiology is very different from that of pathological microbiology, which is mainly concerned with study of single species of micro-organisms. The only pure microbial cultures associated with soil appear to be those inside the nodules of the few Orders of host-plants able to 'fix' atmospheric nitrogen by symbiotic partnership with particular species of micro-organisms. These nodules are plant-physiological rather than soil phenomena; thus it can be said that everything in soil microbiology proper is a matter of equilibria between microbial species. For that reason alone, the teaching of soil microbiology calls for an approach quite different from, and more truly fundamental than, that of instruction in pathological microbiology – upon which teaching of soil microbiology has been mainly and somewhat unfortunately based.

Soil microbiology, then, has two guiding principles. One is the principle of equilibrium common to all ecological studies, yet manifested more delicately in soil microbiology than in any study of the larger organisms, and perhaps most delicately of all in soil bacteriology. The other is expressed in the everyday knowledge that water always finds its own level, or runs downhill. This is fundamental to life on land.

It is possible to put the principle that 'water always runs downhill' into apparently more scientific dress by invoking kinetic energy. To do so, however, brings no biological clarification, because the soil microbes are not influenced by kinetic energy as such. What does affect them is the ratio of air to water in the soil spaces: this ratio changes in detail with minute variations in soil, but it is important to grasp the general idea that the air-water ratio varies broadly from top to bottom of a uniform slope in conformity with the tendency of water to run downhill. Hence the aeration varies along a slope, as well as varying in time at each point in the soil of the slope. These changes result in population-shifts – a series of fluctuations in numbers and species – wherein the important thing is not simply the numbers of individuals of each species, but the balances between species. Considered singly, these balances are fugitive; they depend upon many other things besides aeration: and notably upon the supply of organic matter suitable to and available for each species in what may be a long and involved successional chain.

The intensity with which the linked microbial inhabitants of soil perform their functions of converting the effete matter of the last generation into plant-nutrients for the next depends upon aeration: that is to say, upon the ratio of the extents to which the soil's spaces are occupied by water and air. If this intensity of aeration and of microbial out-turn is high, the production of incompletely-oxidized and unassimilable and possibly toxic products such as hydrogen sulphide, organic acids, and methane, will be negligible. The intensity of microbial activities will vary along a slope; hence the biotic conditions will be more or less propitious for oxidation or reduction in accordance with the drainage and the summation of water movements.

That drainage is of the first importance as a practical agricultural measure is known to every agriculturist; consequently it may be objected that this argument about water reveals nothing new. Yet, it would be difficult to show that any critical examination of the foundations of soil microbiology has been made. It is known among sewage microbiologists, however, that streams are self-purifying so that their waters become progressively 'purer', in the sense of becoming less polluted and better oxygenated, the further they are below the point of discharge of an effluent containing polluting organic matter, *i.e.*, matter capable of being oxidized and therefore capable of using up the dissolved oxygen in the course of microbial or chemical oxidative processes. Interesting features of the correspondence between soil and stream are that both depend upon oxygenation and flow of water, but, whereas the flow of a stream leads to increasing aeration downhill, in soil the motions of water tend to decrease aeration towards the foot of a slope, and hence to a lessened intensity of microbial oxidation and to a slower turning of the wheel of life.

Since life on land depends – the results of sea-fishing excepted – upon the activities of soil micro-organisms, and since the locus of the point of balance of soil micro-organic populations depends upon the water-air ratio in soil, it may be said that the fundamental principle of soil microbiology is astonishingly simple: it being the fact that water always flows downhill. And, among the branches of ecology, soil microbiology is that one most intimately and exactly concerned by the flow of soil water.

Not everyone will accept the view that water and gravitation are indeed the fundamental principles, or elements, of life. It can hardly be gainsaid that no science has a more simple and elegant unifying principle than that fundamental to soil microbiology.

* * *

WHILE this essay is primarily concerned with soil microbiology, it may be taken as a plea for a re-examination of soil classification in the light of first principles. If it be agreed that any natural soil is the product and expression of all the factors: parent material, climate, vegetation, topography, and time: then the ecological aspect becomes paramount, being the only one which is dynamic.

The factor parent material is broadly an accident of locality, or an incident in space; though, because of such facts as that parent material may have been glacier-borne, and because frost action does not occur in tropical climates, parent material is not entirely divorced from climate and time. The factor climate includes regular progressions of latitudinal and altitudinal features; also a set of precipitation régimes. These last do not vary simply with climate; they vary also with such unclassifiable features as continentality, and presence or absence of orographical barriers (*e.g.*, the Rocky Mountains), all of which might be called topographical. But these major topographical features are each unique. The only truly topographical feature which can give rise to a set of soils locally diverse yet susceptible of being matched elsewhere and hence of affording a worthwhile classification, are the minor incidents of topography exhibited by the smaller slopes and by micro-relief, from both of which the major climatic effects of altitude, orography, and continentality are excluded.

It seems that the uniqueness of the orographical and continental influences to which European Russia is exposed accounts for the fairly regular variations or correlations of its climate with latitude; these variations occur in a sequence matched in hardly any other large region, if at all.

Soil classification, in a systematic sense, originated in Russia in the last century. So, although something which can be called a podzol (a Russian word) may be recognized in many humid regions, and a black earth somewhat like a chernozem (another Russian term) may be recognized as a type in other places, the logical way to regard soils is to look upon them as essentially local products and as ecological expressions in which the effects, on the parent material, of the micro-flora, micro-fauna, and vegetation are allowed due weight.

These effects are governed by local relief over quite small areas, and very largely by the movement of water. Since micro-organisms have collectively almost world-wide distribution, the local motions of water do govern their ecology. For the larger plants, however, the governing influence of water-movement upon ecology is seen only in the smaller areas over which free dissemination is possible. These areas may be sometimes almost as large as a continent, though often they will be much smaller. Once it is recognized that distribution and deployment of plants is subject to geographical accidents, the argument for beginning soil classification 'at the bottom' and for working from small regional types towards larger groups becomes stronger, and the case for insisting on Russian-inspired 'great soil groups' as bases for actual soils of a given locality becomes correspondingly weaker.

N. C. W. Beadle[1] has pointed out that, in Australia, species of *Eucalyptus*, *Casuarina*, *Acacia*, etc., form tall woodland under climatic conditions which in North America are compatible only with grassland. If eucalypts and the other flora peculiar to (say) Australia were as free to move as the bacteria are, might not our conception of great soil groups be very different from what has developed?

It may be suggested, therefore, that to descend from a great soil group to local conditions (such as those of a farm survey) is to rely overmuch upon effects of unique and mainly historical chances. A realist approach to soil morphology is to examine the concept of soil groups by criteria derived from soils in the field wherever they may be. If that is done in considering the formation of natural soils, the hydrologic sequence in particular and the roles of water in general will be seen to be as important for soil morphology as they are for microbial ecology.

1. Beadle, N. C. W., *Ecology*, 1951, **32**, pp. 343–345. ['The misuse of climate as an indicator of vegetation and soils.']

CHAPTER EIGHT

Notes on Animal Nutrition

'Come, fetch out the plum-cake, old man!' the Unicorn went on ... 'None of your brown bread for me!' – *Through the Looking-glass*

LIKE any other large animal, man normally requires almost the whole of his nitrogenous intake to be in the form of protein. Free amino-acids, not combined into proteins, are a good substitute nutritionally, but cannot be obtained in quantity except by chemical treatment of proteins existing in meat or vegetables. To sustain an increased population of animals, there must be more protein to feed them with.

There must also be more carbohydrate such as is obtained from cereals and turnips; man likes sugar, too; and must have fat or oil for cooking. The problem of carbohydrate supply is relatively easy, if only for the reason that many plants store carbohydrate in seeds and bulbs and tubers and other structures which are easy to harvest and process, or to eat without processing; oil-production is a little more difficult; but protein-supply is the most difficult of all to ensure in quantity and in concentrated form, *i.e.*, without the protein's being accompanied with too much carbohydrate. All parts of plants contain some protein; the difficulty in eating a diet of seeds or fruits and tubers and the like – those being the parts of plants most easily gathered – is to find enough protein to balance their large excess of carbohydrate and sometimes oil.

Any increase of plant growth results in a gross or over-all increase in the total amount of protein, along with an increase of carbohydrates, also fats. For brevity and for the present the fats and oils can be lumped with the carbohydrates, since both consist only of carbon, hydrogen, and oxygen, in principle. Some of the fat-like substances – lecithins – contain an essential proportion of nitrogen; but these are not a large fraction of the diet, and their main sources are animal products and leguminous plants; so they can be ignored in an introduction.

Probably, protein has always been the main factor limiting human population. Whether one looks upon the wholly or largely vegetarian peoples, or upon the more northerly big nations which principally rely for their protein and for some other nutrients upon the products of animal husbandry, the condition of having, or keeping, more animals and men alive is to be found in satisfying the implications of the phrase 'more protein to feed them with'.

It happens that animals cannot take hold of enough protein if it is diluted with overmuch carbohydrate. This fact is the basis of rationing the organic constituents of foods for farm livestock: the proportion of protein and of carbohydrate must be approximately right, or the animals will not do well, and will not produce the meat or milk or eggs or fat or offspring required of them by us. It is very much (though not wholly) a question of bulk: a stomach is never large enough to hold all the turnips or potatoes or wheat or maize that would supply a daily requirement of protein.

Such mainly carbohydrate foods represent the plants' store of carbohydrate for their use in the next generation. Plants, generally, do not need to store much combined nitrogen for their offspring; for, from a soil no more than moderately propitious for plant growth, plants of all ages and kinds can get nitrogen from the nitrates and ammonia which are unutilizable by animals.

An expression of the need of animals for 'more protein' means that animals need to have a fairly high proportion of protein in reasonably small bulk. Animals need something distinctly different from what the *majority* of *mature* plants supply in making provision for their next generation.

The operative words here are those in italics. *Mature*: any young plant, or new growth of a plant, is at least fairly rich in protein. *Majority*: herbaceous leguminous plants are rich in protein in most of their parts at all stages of growth. Richest in protein while young, and then not necessarily richer (reckoning in percentages of tissue) than any other plant such as a true grass at a similar stage of growth, the leguminous herbs retain throughout their life a greater richness in protein than any other kind of plant of similar age; and this greater proportion of protein extends to the seeds of leguminous plants. The richness of legume seeds, such as beans, in protein (and sometimes oil) is equalled only by some kinds of nuts.

For general practice there are only three ways of obtaining protein in small bulk and high concentration, either for eating as the main constituent of diet or for supplementing the deficiencies of protein in carbohydrate-rich foods. These are:

Living carnivorously;
Eating any kind of young leafy plant;
Eating leguminous herbs at any stage of growth (including their seeds).

A fourth method is by industrial manipulation of foodstuffs so as to extract the protein, or to concentrate it by removing other constituents (as when oil-seeds are pressed).

Dietetically, there is something to be said in favour of cannibalism. If dog eats dog, the eater gets amino-acids of the kinds it most needs and as nearly as possible in the right proportions. Vitamins may be deficient. . . . but need we go on?

Living by eating animals or animal products of the land must either mean being content with what the soil naturally supplies, or else raising the output of animals. In any event, the consequences and implications must be faced; including the implication that to live carnivorously is to depend upon the other two methods of provision of nutrients through green plants. For man to eat fish or to give fish-meal to livestock is a legitimate method of supplementing protein derivable from the land, and of 'balancing' an excess of carbohydrate (say, from cereals).

The second alternative is adopted by animals at grass or browsing young growth of shrubs. In a modified form it is adopted by man in eating some vegetables harvested in a young state; and such artificial practices as grass-drying depend upon it.

As a means of feeding livestock this second alternative has serious limitations. It boils down in practice into use of grassland, whereon the principal vegetation may consist of true grasses, but is more likely to be a mixture of true grasses and leguminous plants such as clovers. If clovers or other leguminous plants are dispensed with, or are not encouraged, then either the yield of herbage – and consequently of animals – is apt to be low; or a need arises for special methods of management involving more skill, attention, capital, and expenditure of fuel than most farmers care to incur or to think about. The production of artificially dried grass is an example of these special methods; though it, or other methods of intensive utilization of 'grass', need not be accompanied by elimination of clovers. It has been estimated that in British conditions the drying of grass to secure a high-protein product demands the consumption of about as much fuel as would give heat equivalent to that obtainable by burning the entire crop. In climates more favourable to the growing of leguminous crops of hay and edible seeds than the British climate is, there are more economical ways of securing protein than that!

There is one method, approaching the large scale, of applying the second alternative to provision of human food. This depends upon cultivation of the cabbage tribe – including kale, brussels sprouts, and cauliflower and broccoli. These are harvested immature; and both leaves and 'heads' are fairly rich in protein. As with the true grasses and other leafy herbs, the growth of these plants, and consequently the bulk, can be appreciably assisted by use of nitrogenous fertilizer (if other conditions are also made suitable). By cutting the cabbage young, it offers practically the only way of converting nitrogen from fertilizers into an increased mass of protein suitable for direct human consumption. The protein seems, unfortunately, to be only second-class; yet broccoli with added fat, or still better, cauliflower *au gratin* comes very close to providing a balanced meal in a single dish. It may be regretted that the possibilities are so limited.

Leaving the third alternative for the space of a paragraph, concentration of protein by squeezing out the oil from seeds and nuts may be briefly mentioned. The seeds may be leguminous, and therefore (like the nuts) initially rich in protein and in oil, or non-leguminous, like cotton or linseed. The concentration consists of removing the oil by pressure, thus raising the percentage of protein in what is left. Leguminous seeds, such as peanut (ground nut; *Arachis hypogea*) and soya bean (*Soja max*), give the richest protein-concentrates available in ordinary commerce. It has been proposed to extract protein from leaves, but no large-scale application of that has been made.

The crux of feeding animals is actually an encouragement of leguminous plants, because they are the source of the most ordinary forms of protein obtainable from the land. Man is an animal. Leguminous plants of all kinds – including shrubs and trees for once – have a further function to fulfil in agriculture through their special aptitudes as nitrogen gatherers, whereby they assist in maintaining a relatively high level of combined nitrogen in the soil, at the same time promoting the formation of organic matter, and thus improving the soil's structure and workability. This effect, or congeries of effects, of leguminous plants upon the soil is usually best made use of when the legumes are grown in association with other plants: either actually mixed with them, as in grass or in the association of leguminous 'shade-trees' with cacao in the West Indies; or in a rotation. Either type of association of legumes with other crops ensures that the other crops benefit to some extent from the extra nitrogenous matter bequeathed to the soil by the legumes.

The classical English or 'Norfolk' four-course rotation lasts four years: (1) turnips (with dung); (2) cereal (sown mixed with 'annual' red clover or with red clover + grasses); (3) red clover (or clover-grass) hay; (4) cereal. The effects of the clover in giving nitrogenous manuring to the soil and in improving its structure are appreciable in three of the four years; in the remaining year, dung supplies nitrogen and other nutrients, and organic matter, from outside the field – though that dung must be credited to the same or similar fields which have provided turnips, clover, and other food for the beasts. The effect of the clover in feeding the livestock through the nitrogen attracted by the clover from the air is thus plainly discernible in this simple rotation.

A similar principle in relation to feeding of animals and raising the nitrogenous and physical status of the soil *for several years* by cultivation of legumes is the core of mixed husbandry, and therefore of the First Agricultural Revolution which began in the eighteenth century, and of the Second, which took shape in our own century.

Whether man eats pulses or other leguminous products directly, or prefers to eat meat or cheese or butter derived from clovers and grasses which he cannot himself eat, the food problem will not be solved by encouragement of mere photosynthesis at large. The food problem demands discriminate attention to the question of promotion of protein in the first place, mainly through conscious and deliberate encouragement (I would say *cultivation*, if the word had not been pre-empted!) of leguminous plants. That will have the effect of soil-improvement which will ease the rest of the problem. Production of carbohydrate and oil from cereals and non-leguminous plants is relatively easy, given the land *and* the ability to cover it for a least part of the time with leguminous plants alone or with legumes in mixture with grasses.

If cereals and other carbohydrate foods are mainly grown, the problem of obtaining protein remains, or is aggravated, for both man and livestock. If the problem of extending legume-cultivation is solved at least partially, any question of balancing the due proportions of human nutrients can be regarded as solved, or as practically not arising: at least to the extent that total food supply is adequate for a given world population.

Every effort of man to supply himself with food must be in accordance with the principles of nutrition and with the cycle of water, carbon, and nitrogen. He must work always with full understanding of what he is aiming at and doing. Else he will only stub his toes, and some of his efforts will not only be in vain, but painful; and an effect opposite from what was intended may be brought about. The production of such an unmeant effect is not uncommon as a result of random or chance or undirected efforts. In fact, the odds for or against such a result from a random process are about even: the wrong effect will be produced about as often as the right one.

(The African ground-nut scheme *might* have succeeded. To a scientist conversant with probabilities, the stakes of several tens of millions of pounds seem a little high at such poor odds.)

I am not competent to discuss prospects of greater supplies of fish, so I restrict myself to pointing out the very great importance of encouraging the biological fixation of nitrogen. That is our only substantial hope from the land. In spite of all technical advances, it remains true that bacterial fixation of nitrogen, by legume-nodule bacteria in partnership with leguminous herbaceous crop plants, is the chief source of protein from land for man and other animals.

This means, conscious or unconscious exploitation of certain groups of bacteria to get more food for people. Only truly microscopic organisms, all of which are soil-dwellers, can make direct use of air-nitrogen either for themselves or in such a way as to make air-nitrogen available to green plants. In connexion with production of notable amounts of bulk protein on land, 'green plants' can mean hardly anything but the herbaceous legumes, such as clovers, lucerne, and beans.

There is no substitute in sight for the legumes as providers of protein in forms now understood by man and acceptable to man as food.

It is peculiarly difficult to convince stay-at-home Britons of the importance of legumes. It is probably useless to mention that most New Zealanders are clover-conscious; or that a large proportion of farmers in Australia and the United States are well aware of what leguminous crops and leguminous pasture-plants mean to better farming and higher outturn of animal products. We in the United Kingdom depend for a large part of our rations of animal produce upon what has been raised on the clovers and other legumes of New Zealand, Australia, Denmark, and other countries where legumes are *recognized* as the chief providers of nitrogen for animal consumption and for conversion by farm animals into protein; but, not being really interested in agriculture or in food-production, and tending to think that food is something which normally comes in a ship from a place a long way off, we mostly don't know the origins of food, and don't care. We have learnt that milk doesn't really come out of a bottle, but actually comes from an animal; but beyond that we have not progressed appreciably. If there is any widespread impression about food-production, it probably runs on these lines: because protein contains nitrogen, and we can now fix nitrogen from the air by synthetic means, the fertilizer problem has been solved, at least potentially; and the solution to the food problem is in sight as soon as we can build more nitrogen-fixing factories.

That is incorrect. There is still – I repeat – no substitute for legumes.

In the United Kingdom there is no legume of high importance beyond the clovers: annual, red, for hay (*Trifolium pratense*); and perennial, white, for grazing land (*Trifolium repens*). Clovers are apt to be confused with grass, or to be lost from sight among the herbage of grassland. There are few more strikingly beautiful sights than a field of crimson clover (*Trifolium incarnatum*): but how rare that is! A Briton passing through Lombardy – where 'grass' in our sense does not grow – can hardly avoid noticing that about every third field is of lucerne (*Medicago sativa: erba medica* the hay of the Medes, and plausibly the grass which Nebuchadnezzar ate. *Daniel* iv, 33). This does duty for grassland in producing hay and protein.

We Britons are so used to seeing beans only in the white sorts which come from overseas in packets and tins that we are not ready to conceive that, in the lands from which their seeds reach us, bean plants do in reality grow in quantity at someone's doorstep. Peanuts have no pleasant associations for us, though they have long been the chief crop and principal export of our least-known African colony; and as for soya-beans, we associate them only with trivialities like the less happy memories about ingredients of war-time sausages. Yet soya-beans are staples of food and industry in several of the most important countries.

We are never at home with the podded plants except as an occasional fresh vegetable. We do not live by legumes (or so we think); certainly not upon them. They are an extra to living; at worst a joke, or, like soya-bean flour, a foreign intrusion to be shrugged off as soon as possible; else, most likely, a pale mass in a tureen as a despairing accompaniment to potatoes when our only other vegetables – cabbage, cabbage, and cabbage – cannot be obtained.

If the North Italian scene looks singularly un-British, it may be worth remembering that the insistence of agriculture almost everywhere upon lucerne or another leguminous herb (even indigo is not despised in parts of Africa and China) is not just one of those foreign ways which foreigners so oddly have. It keeps them alive.

In the world of legumes it is we in the United Kingdom who are the strangers and odd-men-out: to be pitied for not being generally able to grow any legumes except clover, and not being able to see even that when it is literally beneath our feet. That is, if we are grown-up. Children acknowledge 'eggs-and-bacon' (*Lotus corniculatus*) for its pretty speckled flowers so aptly indicated by their name for it; and there is the shamrock. And sweet-peas . . . but we are back in the town.

As a corrective, I propose this exercise for the urban student: ascertain how, and at what times, a clover goes to sleep. (The results need not be reported to me.)

It is true that a very small share of the earth's nitrogen is required by any crop, leguminous or other. A crop of a type that is rich in nitrogen and is harvested only once contains only about a hundred pounds per acre at its maturity. Sometimes this figure is exceeded; but an average crop contains much less. Atmospheric pressure per square inch is 15 lb., of which nitrogen accounts for twelve. Thus, a crop on an acre utilizes during a year only as much nitrogen as exists over a few square inches of the surface it occupies.

This estimate includes the nitrogen derived from rain and soil by leguminous and non-leguminous plants alike (*i.e.*, plants which respectively can, and cannot, make direct use of air-nitrogen); it is not meant as an estimate of the extent of biological fixation by the legume-nodule bacteria. However, as the subject has been raised, it may be as well to dispose of it. There is no certain way of making an apportionment of nitrogen biologically fixed by legume-nodule bacteria and other micro-organisms. In temperate climates the rain contributes about 4 lb. of combined nitrogen per acre per annum; and the contribution of species of *Azotobacter* and other bacteria and micro-organisms which fix nitrogen while living free (*i.e.*, without being in partnership with plants) is about as much, possibly. The rest which comes from the soil is as ammonia and nitrate produced by the soil bacteria in the final stages of decomposition of nitrogenous material – animal and plant proteins, urine, manures and nitrogen-containing fertilizers (if not already nitrate) – added to soil.

Nitrogenous fertilizers are not usually added in large quantities when a leguminous crop is being grown or is about to be grown; so we can make a very rough estimate of about 30–50 lb. of nitrogen taken up per acre per annum from the soil and rain alone, by any kind of crop. Any remainder must be credited to the activities of the legume-nodule bacteria in symbiosis with their host plants. Without obvious dishonesty, the contribution of the legume bacteria + legume host plants to utilization of nitrogen by those legume plants may be taken as being generally appreciably greater than the total of nitrogen derived from soil and rain by the legumes or by any other kind of crop.

Forty, or perhaps fifty, pounds of nitrogen per acre may be normally credited to the legume-nodule bacteria in association with their hosts; that is to say, the nitrogen available for plant growth, and as protein for animal feeding, is at least doubled when the crop is a leguminous one grown under good conditions, as compared with what is obtained from a non-leguminous crop grown under the same equally good conditions of soil treatment and manuring. The range of annual gain of extra nitrogen from growing a leguminous crop may be put for most legumes – that is intended to mean, those which are harvested once, like soya-beans or peanuts grown to maturity, with the seed as the principal form of produce – at between about 30 and 80 lb. of nitrogen per acre.

Fodder crops such as lucerne, which can be cut several times a year to give a ton or so of hay at each cutting, give a much greater yield of biologically-fixed or 'extra' nitrogen in the green tops or hay alone, as well as leaving a large poundage of nitrogen in the soil in the form of roots; and the miscellaneous soil microbes make that root-nitrogen available as a form of manuring to subsequent (usually, non-leguminous) crops during several years. Ignoring such gains in root-nitrogen (= ultimately soil nitrogen), the amount of nitrogen bacterially fixed through growing of a good crop of lucerne may amount to 100 lb. per acre or more each year while the crop lasts.

If we take the total of nitrogen obtained by a nitrogen-rich crop from all sources as 80 lb., and assume all to be protein, or to have equal value to protein for feeding purposes, we should get about 480–500 lb. of protein per acre; or about four hundredweights and a half. That is not far from the *maximum* yield of protein obtainable *on an average* by ordinary agriculture from an acre. As has been implied, lucerne and some other leguminous crops are capable of giving appreciably more than that annual yield of protein. Such high yields as those which can be obtained from highly-farmed lucerne are associated with crops of which a portion, or none, of the protein is in foods suitable for direct human consumption. Yields of total protein from pulse crops may be, exceptionally, similar; but much of the nitrogen of those crops is in stems and leaves, which may be used as animal fodder or for immediate incorporation in the soil, but cannot be eaten by man.

Taking all considerations into account, the yield of vegetable protein eaten by man that is obtained by ordinary agriculture is apt to be less than a ton per five acres of leguminous crop. A heavy crop of once-cut hay may yield about the same for animals: providing that there is a good proportion of clover or other legumes among the grasses.

On account of losses in conversion, yields of animal protein per acre per annum are much smaller. It is an old saying that it takes as long to build a bullock as to build a battleship. The maximum animal crop that can be sustained on good land is about one cow or bullock per acre of good grass, or roughly the number of such quadrupeds will be about a quarter or a fifth of the acreage of the whole farm.

People sometimes lament the immense herds of bison which once roamed the North American prairies. These masses of animals spectacularly appeal to the imagination, so that one might think that there has been some declension since the bison disappeared. The venue has shifted, but that is all. Millions of cattle are now reared on the range lands, and more millions are fattened on legumes and cereals in the Corn Belt to the east, though the former prairies are mostly given up to continuous wheat. What is lost from sight in recalling those millions of bison is the vast acreage over which they roamed in what are now two countries. If they were to be thought of as spread out over their original lands, it would be realized that the bison were actually very thin on the ground, and the ratio of land and animals was of the order of acres per animal (as it still necessarily is in the

Scottish Highlands); not animals per acre, or an animal per acre.

There is, in recalling those bison herds, a very striking point about the necessity of abundant earthy materials for animal construction. The bison roamed mostly west of the 97th meridian; that is to say mostly west of the Missouri rather than west of the Mississippi, and mainly west of the State of Iowa and the rest of the U.S. Corn Belt now bursting with cattle and pigs. The railway bridge at Omaha bears on its eastern portal a figure of a bison, to show that the traveller is about to enter the bison country. This region from the Missouri to the mountains is dry, and unfit for growing any crop except wheat; only eastwards is there enough precipitation for mixed farming to be undertaken. Because the bison country was not much rained upon, its soils retained their lime and phosphorus, and still do so, and still will keep them so long as the land is not irrigated 'to make it more productive'. (See page 234.)

Only where lime and phosphate were abundant in the soil did the bison flourish. The eastern part of North America was covered with trees – not grass – in balance with the poorer soils produced by the humid or semi-humid climates; from those soils, and especially on the coastal fringe, much of the mineral nutrients fit for building bone and milk had been washed out. This eastern part of North America produced no wild animal larger than a turkey.

The moral is that lime and phosphate in quantity must be provided naturally or by art wherever animals are kept in quantity – even if the density of stocking is less than one per acre. Lime and phosphate (and the other earthy substances needed by animals) are easier to supply than a corresponding amount of protein; but a high rate of animal production implies both a balance between protein and carbohydrate and a balance between those organic substances and the essential inorganic, earthy ones. Protein cannot be induced to grow in vegetable form for animal consumption without a good sufficiency of lime and phosphate in the soil. We will look further into that proposition in the next chapters.

CHAPTER NINE

Down to Earths

'Calces o' fossils, earth, and trees;
'True Sal-marinum o' the seas;
'The Farina of beans and pease,
'He has't in plenty;'

Robert Burns, *Death and Doctor Hornbook*

THE average yield obtained by cropping the soils of the 'Western' or technically-civilized countries is at least 30 per cent more than the bulk of vegetation that could be obtained from the soil if it bore the same crops, but were only tilled to enable those crops to be sown. To put it in another equivalent way, the human population sustained by the soils of those peoples is at least 30 per cent greater than the population which could be fed by the crops if the soil were cultivated but unnurtured, and if no importations from overseas were admitted. The increases in bulk of vegetation and of population now obtained over what the soil would naturally sustain annually are far greater; but a figure of about 30 per cent will do to go on with and to stimulate thought about responsibilities towards the soil now and for the future. You may glance at the diagram on page 209, meanwhile.

About a third of us, possibly more of us, could not stay alive but for reasons adumbrated in what follows.

When any kind of crop is cut for feeding to man or animals, or is grazed where it grows, or is removed in the process of mowing the lawn, something is taken away. This something includes:

(1) the elements caught by any kind of plant out of the air: the carbon and oxygen of carbon dioxide, and the hydrogen of water;

(2) the elements taken by any plant out of the soil: these will include nitrogen (as nitrate or ammonia), phosphorus (as phosphate), and others liberated from previously-living matter by the actions of soil microbes; also hydrogen and oxygen from soil water; and combined nitrogen from rain and free-living nitrogen-fixers;

(3) the metallic and other elements belonging to the earth – as distinct from the air – including calcium, magnesium, phosphorus, etc., set free from bones and other organic structures through the agency of soil microbes, as well as the same and similar elements made available by slow solution of the rock minerals;

(4) for leguminous plants only (in practice, that is, and neglecting the finer distinctions): nitrogen gained from the air through the association of leguminous plants with their nodule bacteria.

It will be seen that, of these losses brought about by cropping, two – (1) and (4) – are borne by the air. The gases of the air are truly inexhaustible, in relation to the bulk of even the most intensively-grown vegetation. What is more important, possibly – if not to the plants, then to our argument – than the inexhaustibleness of the three atmospheric gases is the fact that they are all completely available to every plant suitably equipped to deal with them, up to the limit of growth permitted by the reigning climatic conditions such as sunlight and season. All green plants can make free use of the elements specified in (1); all leguminous crop plants can make free use of nitrogen as specified in (4).

The limiting factors to plant growth and crop production under any given conditions of climate and season are in practice the non-metallic substances water, nitrate, and ammonia and the elements phosphorus, calcium, magnesium, sulphur (as sulphate) and some others, all of which are derived by non-leguminous plants wholly from the soil.[1]

1. It is probable that plants take up small amounts of nitrogenous compounds absorbed in rain or dew deposited on their leaves; if this kind of accession of combined nitrogen is thought to be material, an appropriate correction can be read in. It does not affect the main argument.

Whether or not they bear root-nodules containing the invaluable nitrogen-fixing bacteria, leguminous plants behave exactly similarly. They depend for their nutrition upon the soil, and upon what goes on in the soil around their roots. Like non-leguminous plants, legumes require the products of rock weathering and of general soil-microbial activity, so that their general growth, or extent and activity of growth, depends on, and is limited by, what they can get out of the earth.

In view of the unique ability of leguminous crops to profit from nitrogen of the air, this is worth repeating in refined form. Nitrogen of the air is illimitable in quantity and availability; leguminous crops can fix nitrogen from the air; but they can do so only to the extent that both their growth and the nitrogen-fixation which goes on within them is assisted by ample supply of the earthy elements phosphorus, calcium, magnesium, sulphur, etc. To leguminous plants, as to others, it does not matter how or by what route the elements mentioned in (3) and (4) are supplied, so long as they are in compounds which the plants can take up.

Losses (1) and (4) do not matter, nutritionally; nor do losses of rain-nitrogen or anything else derived wholly from the air: they can be replaced. Losses of the *earthy* elements mentioned in (2) and (3) matter very much indeed, because in so far as those earthy elements are taken away from the soil by cropping or any other mode of away-carriage, the limiting factor to growth is itself made smaller.

Nitrogen and its compounds is, clearly, in a special relationship to plants. Some of it is combined nitrogen (ammonia and nitrate) in the rain and soil, and is available to *all* green plants. The rest, if not already combined in plant and animal tissue, living or fossil, is in the air as gas; and as such is usable only by the legumes equipped with nodules containing active nitrogen-fixing bacteria. We will come back to that in detail later.

Besides losses induced by cropping, and most notably by cropping for consumption off the farm, there are other losses in most, if not all, soils. Especially in humid climates, where the losses may be very appreciable in the course of one year, there are losses from the leaching or washing-out effect of rain. Rain takes away nutrients in solution, or as fine mud. The rate at which soil goes to the sea depends on amount, form, and intensity of precipitation, and other climatic factors; also upon type of vegetative cover such as natural grass or forest, etc. (that being in nature not independent of climate), or kind of crop (and especially the degree to which it covers the soil). Calcium, or 'lime', suffers severely from being washed out by rain, because its commonest soil forms (carbonate and sulphate) are appreciably soluble in the carbonated water which exists in soil.

For introductory purposes we can neglect the losses of the generality of the soil as experienced under conditions of good, conservative, farming as practised in the United Kingdom and the rest of north-western Europe, and a few other regions of the world. For such sensible conditions of farming and a not too aggressive climate, it may be assumed that the small annual losses of the body of the soil are compensated for by an equal small annual deepening; so that little or no effective change results over a lifetime or two. (Where soil erosion is severe, that assumption of equilibrium is shriekingly untenable.)

We have, therefore, the natural losses inescapable owing to the exposure of land to the atmosphere. Under wild conditions these reach equilibrium with the natural cover of trees or shrubs or grass, and with the natural fauna, including savage man. The savage limits himself – or did limit himself – to a natural equilibrium with the soil; limiting his numbers by head-hunting, warfare, cannibalism, infanticide and so on to what can be sustained by the natural or almost natural sources of food; or, if he does not limit his numbers by such human techniques of restriction of population, he has them limited for him by famine or disease or catastrophe.

Under agricultural conditions the same natural losses of plant nutrients from the soil occur, but they are greatly augmented by the removal of natural plant-cover which is the essential, and essentially artificial, preliminary to agriculture. The native trees, shrubs, or grasses are removed; and the soil is broken by spade or plough. This intensifies losses of nutrients, and of natural soil organic matter. The keeping of animals, if accompanied by return of as great a proportion as possible of plant and animal remains to the soil by a conservative system of husbandry, lessens the increase in losses of nutrients above what prevails under wholly natural conditions. Of British mixed farming it can be said that farm animals have replaced trees: not quantitatively in the sense of one animal for one tree, but as the principal agent of plant-nutrient conservation, now that the original forest cover of the islands has gone. The loss of plant nutrients, even in Britain, is still severe.

Losses of nutrients from the soil under agricultural conditions are at a practical minimum when subsistence farming is practised. Subsistence farming means that everybody lives on the land and by the land, with few or no imports and hardly any exports. It necessarily connotes a low standard of life for those who practise it; and, rigorously interpreted, it means that no towns can exist, consequently no collective civilization of the industrial and learned types able to be fostered by city life. In fact, of course, there are always some towns, but the proportion of urban civilization can never be large if subsistence farming is to endure; unless the whole of the refuse and excreta of the town populations is effectively returned to the land.

If a close approximation to subsistence farming exists – as it does in parts of Asia now, and as it did in Europe a couple of hundred years ago – the system can last indefinitely, provided that the natural checks of disease and famine are allowed to operate, or some other check comes into operation, to keep the population on the land down to numbers which the land can support. The continuance of the system implies: little leisure (because the back is always bent either in tilling the soil or in collecting effete material to return the nutrients to it); a bare minimum of food; sordid wretchedness, and misery.

If subsistence farming is departed from to an appreciable extent to supply towns nearby, or to export farm products to a great distance – there being little or no return from such exports – one of two things must happen in all but the most favourable and rare conjunction of circumstances. Either the land must 'run down' in condition; or there must be an accession of nutrients from outside the soil. Assuming that, as always in the most highly-developed agricultures supporting the densest populations, leguminous plants are the centre around which the rest of cultivation revolves, there is no need to make good the losses of nitrogen from export of meat, milk, tubers, grain, or any other farm product; but in all circumstances *under a humid climate* there must be an accession of *earthy* materials such as lime, phosphate, and potassium, *if the land is not to become run down.* This addition of *earthy* materials is the *sine qua non* of continued production of food under conditions wherever land not naturally abundantly rich in the earthy plant nutrients is used to feed a population at a distance from the land.

The extent to which these earthy materials, phosphate, lime, potassium (and a good many others in smaller proportions), must be supplied varies with the size of the gap between what is yielded annually by the soil and what is removed from the soil by natural agencies and to meet the demands of the distant population for protein, and for minerals in the forms contained in bones, milk, meat, grain, pulse, and every other kind of soil-derived food.

In view of what has been just read about the absence of need to replace biologically-fixed nitrogen, the reference to protein will probably seem contradictory. I therefore simplify.

Earthy materials (let us call them, for brevity, 'minerals'; and understand by that term particularly calcium or lime, phosphorus or phosphate, and potassium [salts]) must be supplied to make good:

(*a*) the natural losses of earthy materials brought about by leaching by rain falling on cultivated land – leaving aside the questions of the artificiality or degree of closeness to naturalness of farmed land; together with

(*b*) the losses of minerals sent to town or overseas.

Supplementation to that extent is the bare theoretical minimum allowable if the land is not to run down. If the productivity of the land is to be *increased*, in terms of either the crops or livestock, or both, to be raised on a given area, a further addition of minerals must be made. This further instalment of minerals above what is required to keep the soil at its lowest effective level of food production under 'farmed' conditions must be made not just to replace the extra minerals that will be taken from the land in animal and plant crops of all kinds, and what will be lost from the soil, but *to sustain an increased level of nitrogen-fixation* by leguminous plants.

This extra nitrogen-fixation results in extra protein. The equation is simply (if obscurely for the present):

minerals = protein

One way of approaching an understanding of this is to see that it would be useless to send extra animals to market, and to use additional minerals to provide those animals with bone (either solid, or in solution in milk or eggs) unless the extra skeleton can be clothed with protein flesh.

A corresponding argument applies, though with diminished quantitative force, to every other kind of crop, including such protein-poor products as maize. It does not apply at all to certain products of crop-processing if the unmarketed residues are returned to the land: for instance, sugar and oils, which, if pure, contain only carbon, oxygen, and hydrogen, and thus might make no net demand upon the soil minerals or soil nitrogen if a special kind of agriculture were rather improbably devised to produce them alone. However, the extent to which crops yielding sugar and starch and oils can be grown is limited nutritionally, and quite irrespective of agricultural considerations, by the imperious need for protein to balance them.

The other argument for earthy materials as the plant-nutritional supports and props of protein-production in bulk (*i.e.*, by legumes) is that the leguminous crops must be well supplied with phosphate, calcium, and potassium to meet their double need for nutrients: firstly, as plants; secondly, as hosts to an active population of nitrogen-fixing bacteria. In humid climates, at least, there is a third demand for calcium as lime to bring the soil to physico-chemical condition of adjusted acidity and carbon-dioxide balance in which the temperate legumes flourish as plants and as nitrogen-fixing organizations. This physico-chemical requirement for lime is large; in Britain it is of the order of a ton or two per acre every few years; and incidentally, as it were, by meeting it the purely nutrient requirements for calcium are also satisfied. The physico-chemical requirement of the soil for lime is intimately bound up with the ordinary and normal activities of those soil bacteria which, convert non-living nitrogenous matter into ammonia and nitrate. Though important, and quite germane to the matter of nitrogen-fixation by legumes, this question of demand for lime by the ammonifying and nitrifying bacteria can be left for separate discussion (page 140).

CHAPTER TEN

Feeding the Plants

> Many able chemists have also ranged among calcareous earths, those which are procured from vegetables and from bones by combustion. – *A Manual of Chemistry*

AIR and water are all very well, but are no more than a beginning. There are several views about the ways plants take up their 'mineral' or earthy nutrients from the soil, but for our purpose it will do if we regard the roots as taking what they want out of the soil solution – meaning, the soil water and whatever is dissolved in it. Like the composition of the soil's microbial population, the composition of the soil solution is normally in a constant state of flux: it is responsive to changes in the atmosphere, from which it gets rain and air, and to which it yields water and gases. If the soil solution is not in a state of flux – if it is not subject to a more or less rapid series of upheavals and flows of changing equilibria – then it is stagnant; and in stagnant conditions of soil water and soil air no plants will grow, nor any microbes live, except the sulphate-reducers and other bacteria which can get along without atmospheric oxygen.

The composition of the soil solution is in principle minutely sensitive to change (as, for example, change brought about by a small withdrawal of a nutrient by a root; or addition of oxygen from outside; or addition of carbon dioxide by respiration of soil microbes and plant roots). However, the changes occurring in the composition of the solution during a short but measurable time, say an hour or so, are usually small; even when a big external change happens, such as a fall of rain big enough for it to be expected to dilute the soil solution appreciably and to lessen the concentration of the dissolved nutrients at the disposal of plants.

If there is a heavy fall of rain, there is very little change in the *composition* of the soil solution of most soils over the area, or volume, of soil actually wetted. What happens is, roughly, that the incoming water – from whatever source it comes – tends to displace the tenant soil solution to an adjacent level of drier soil until that soil, too, is saturated with soil solution. The fresh water almost immediately dissolves what it can out of the soil reserves; and, if those soil reserves of nutrients and other substances are both ample and in a form ready to be dissolved, they dissolve so that the rain or other water quickly takes on almost the composition of the original soil solution. The only important substance which can be washed out of soil from one layer to another, or from soil into a drain or ditch, without being quickly and effectively replaced from the soil's stores is nitrate.

Nitrate is slowly formed, in good soil, from ammonia by the special groups of nitrifying bacteria (page 141). (In poor soil it may not be formed in quantity, or may be destroyed by other bacteria if it is formed or supplied.) Whether reliance is placed on the nitrifying bacteria to make nitrate, or whether nitrate is supplied as fertilizer, the nitrate is vulnerable to an excess of water. Often, of course, the effect of rain or irrigation is simply to wash the nitrate down to a lower level of soil, so that it is not completely lost; and, once the water in the area first wetted has drained down in part and has thus let air in again, more nitrate may be produced by the nitrifying bacteria in that area, so that the loss of nitrate is made good. This discussion about the fate of nitrate is partly academic; it is sound enough to give an idea of what does happen in uncropped soil and of what may happen in cropped soil, but in point of fact if the soil is covered with a vigorously-growing crop of any kind (including weeds), there is not likely to be much nitrate to be washed out – the roots will have taken up most of it; and, for a time, roots will likely take up nitrate as fast as the bacteria make it.

Except in regard to nitrate (and chloride, which is not very important as a plant nutrient) the composition of the soil solution exhibits a definite resistance to change, as a general feature. This resistance to change in composition of the soil solution will be more or less marked in different soils. It arises from the cushioning or poising or buffering effects of the soil gases – especially carbon dioxide – and of that part of the soil solids which is neither *very* slightly soluble in soil solution (most of the rock minerals, clays, and organic matter) nor *very* soluble in water under soil conditions (nitrate, chloride). The solid substances which govern the composition of the soil solution are, in most British and temperate conditions: limestone or chalk or calcium carbonate; gypsum or calcium sulphate; certain common metallic elements such as calcium, magnesium, potassium, and others, which, together with ammonia, are attached to the soil colloids in such a manner that they can easily be detached and exchanged for other elements, but cannot be readily washed off by water alone; and phosphates, and especially that portion of the soil phosphate which has been recently deposited on and around the soil colloids.

In considering the relations of water in the soil to air in the soil, it was enough for introductory purposes to talk about the soil solution as 'water'. For discussion of the chemistry of plant feeding, we must have regard to what is dissolved in the soil water. Since the gases of soil are soluble in the soil solution, that solution should be looked upon as a dilute solution of various substances in carbonic acid – a solution of carbon dioxide in water. The resultant of solution, and equilibration with water and carbon dioxide of all the substances mentioned in the preceding paragraph, is, for the ordinary types of soil in the United Kingdom, that they contain a soil solution which is largely a solution of calcium sulphate and calcium bicarbonate, with smaller amounts of salts of magnesium and potassium, a little nitrate, and a mere trace of phosphate.

These typify the main nutrients of plants; and, along with

organic matter, they are the main nutrients and sources of energy for the miscellaneous microbes of soil that do the main work of decomposition of dead matter and dejecta of plants and animals: that is to say, those nutrients, along with ammonia evolved during decomposition, form the principal nutrients of all but some specialized types of bacteria; while all living things require combined nitrogen, as well as phosphate and other earthy substances, to build their cells and keep them going while they are alive.

After this brief survey of the dynamics of the nutrients in the soil solution it is desirable to recapitulate a little so as to survey the stock of nutrients actually – not potentially – available at any moment for the feeding demands of plants; available, that is to say, in the sense that they are nutrients on immediate call, either dissolved in the solution or freely exchangeable with it.

These available nutrients include nitrogen as soluble nitrate, backed by a reserve of exchangeable ammonium attached to the colloids while awaiting bacterial conversion to nitrate; sulphate, which is also water-soluble, and is to some extent a product of microbial decomposition of the sulphur-containing compounds of plant and animal and microbial tissue, but is largely dissolved from the soil minerals; calcium, magnesium, potassium, and other metallic elements attached to the colloids (as ammonia is) in the electrically-charged form known as ions (specifically cations or kations: *pronounced* cat-ions); and phosphate in recently-precipitated form scattered very finely divided among the other solids, a small fraction of it being in solution. The phosphorus of phosphate will have been neither oxidized nor reduced; it may have been set free from bones or other organic sources by bacteria, or may have come out of the soil minerals. As with sulphate, these minerals may be either the natural ones, or may be a stock built up by additions of fertilizer.

In most of the productive soils of the temperate regions, if not of the whole world, the nutrient present in greatest proportion in the soil solution is calcium. Reserves of calcium salts exist commonly in the soil as the solid sulphate and the equally solid carbonate. These reserves may be wholly natural, or may have been added to by man. In either form the calcium is quickly available, being replaced in the soil solution as fast as it is withdrawn by plants. Not so with phosphate locked up in the soil minerals or in fossil phosphate added to soil. Unless supplemented by recent additions of phosphate in some form which can be called plant-available, the soil phosphate can be, and often is, a limiting factor in plant growth.

The amount or availability of water, air, any one nutrient, or any other circumstance can be a, or the, factor limiting plant growth under natural conditions. Under wild conditions, the character and quantity of the vegetation adjusts itself accordingly. Supposing water and air enough, supplies of nitrogen and phosphate are most likely to be the factors limiting plant growth and consequently the density of animals.

Unless something is done to the soil to increase the amounts and turnover of nitrogen and phosphate at least, the same will be true of humid soils such as are the basis of temperate and tropical agricultures: in general for such soils, the crops obtainable will be severely limited unless the amounts of plant-available phosphate, nitrogen, and perhaps other nutrients are all simultaneously raised to levels adequate to support a density of vegetation – of cropping, that is – much higher than what is natural to the soil.

In humid temperate climates at least, attention must also be paid to the physical state of the soil, especially as that is reflected in the minute proportion of hydrogen attached to the soil colloids and governing the acidity of the soil and the soil solution. Like ammonium and the metallic elements attached to the colloids, this hydrogen – so important in governing the physical state of the soil in many ways, and in determining the availability of the other nutrient elements for plants – is not present as the element or gas, but in the ionized condition: it has an unneutralized positive electrical charge. Traces of ionic hydrogen are also present in water and in the soil solution. The great reserve of hydrogen is water.

The more appreciable is the trace of ionic hydrogen in the soil solution, the greater the effective acidity of the soil. Ionic hydrogen attached to the soil colloids constitutes, as it were, a reserve of acidity; but that is only a manner of speaking, since the soil acidity is an immensely complex expression of all the soil factors and conditions, so that to refer soil acidity to the soil colloids is hardly more than a convenient way of looking at things.

It is to the hydrogen on the soil colloids, plus the carbon dioxide in the soil air, that we may look for a decision about the point of balance of the quantity and activity of the hydrogen ions in the soil solution, and thus to the effective acidity of the soil and for a high degree of control of the kinds and amounts of nutrients which the plant roots can extract and utilize from the complex of soil solution and soil.

To 'lime' soil – by adding calcium carbonate (as ground limestone, for example) or calcium hydroxide (slaked lime) or calcium oxide (quicklime or 'burnt' lime) is one of the important operations of husbandry in the cool humid regions, where calcium is continually being washed out by rain. The number of pounds of lime or calcium taken up by most crops per acre per annum is almost ridiculously small: about ten pounds by a good crop of wheat (straw and all), a little more by a crop of oats or turnips, and fifty pounds or so by a good leguminous crop.

Most plant nutrients are removed by crops and are required annually by them on a scale of the same order – a few pounds or tens of pounds per acre, if calculated as the element, or (as it is common to reckon) as the oxide of the element. Fertilizers containing roughly 20 per cent of their nominal nutrient (nitrogen or phosphate, say) need be given only at a rate of about a couple of hundredweight per acre, or a little more for generous treatment. 'Lime' – whatever the form chosen – usually requires to be supplied at the rate of a ton or two per acre. This automatically covers needs for calcium as nutrient.

The most modern view about the effects of liming is that its purpose is to bring the soil into optimal equilibrium with the carbon dioxide of the air. The usual or text-book view is that liming controls soil acidity. In practice and on a short view, both these ideas amount to the same thing. The newer view is, I think, more comprehensive and suggestive than the older one, if only because acidity is only one factor in a highly involved set of chemical, physical, and micro-biological balances, and of the resultant between these and plants. Since soil can be regarded as nothing but an equilibration mechanism, any concept which assists us in comprehending the balances in soil and in expressing them in our own terms is to be welcomed.

In whatever form it is added to soil, 'lime' is converted into the carbonate. It is as carbonate that 'lime' exists in naturally calcareous soils. Carbonate of lime is not soluble in pure water, but it is slightly soluble in rain water (a solution of carbon dioxide mostly) and in the soil solution, which is a solution of carbon dioxide among salts. So we have solid 'lime' brought into solution as calcium bicarbonate in small but renewable amounts, the rest of the 'lime' remaining undissolved but ready to be dissolved with relative ease. Also, some calcium and other elements are attached, in exchangeable form, to the soil colloids, from which they can be released to make up deficits arising in the soil solution from any cause, such as uptake by plants. Nitrogen and phosphate are peculiar.

There is no reserve of combined nitrogen in the soil, except what is present in organic remains—mostly recent. The amount and rate of nitrogen supply in combined and assimilable form to plants therefore depend upon the amount and kind of organic matter in the soil, plus what is added in inorganic forms by rain and by man. (If man adds *organic* forms of nitrogen, in compost or dried blood or bones or green manure or anything else, that will merely have come from some other soil site, and will have done no more than robbing Peter to pay Paul; or if he adds it to the site it came from, he will be making no net addition: in fact, no net addition in either case.) There is only *one* way of augmenting the effective stock of organic matter in the soil so as to increase appreciably the stock and turnover of nitrogen available to and assimilable by all kinds of plants: that is, by calling upon leguminous plants to bring down nitrogen out of the air so as to infuse air-nitrogen into organic matter.

The process of cultivating leguminous plants tends to raise the soil's stock of combined nitrogen within organic matter of legume roots (and of the tops and other residues ploughed in, as well as of the residues of additional animals sustained by the extra fodder protein). The percentage increase of soil nitrogen may not be very big; and it is important to recognize that, under many exhaustive conditions of cultivation and in what farmers call 'hungry' soils, any mode of addition of nitrogenous organic matter – whether as farmyard manure or compost or the growing of a leguminous crop with or without grass – may do no more than prevent the proportion of organic matter in the soil, and with it the amounts of total and available nitrogen, from falling.

However, there is no alternative to the cultivation of leguminous crops for increasing the general stock of combined plant-available nitrogen in the soil, or for sustaining that stock in the majority of circumstances. Augmentation of the stock of soil organic matter and consequently of nitrogen which the miscellaneous 'decompositional' bacteria, and their ultimate allies the nitrifying bacteria, can put at disposal of crop plants may not go very far; for, to whatever level the stock of soil organic matter is raised by cultivation of leguminous plants or any other device, that level will be artificial: it can be kept up only by continuous human effort.

Whatever man does towards raising the level of organic matter and available nitrogen in the soil, his efforts at striking a higher balance will be resisted and opposed by the soil microbes and other natural agencies. These agencies will co-operate to destroy and remove any unnatural excess of organic matter, and to restore the proportion of organic matter to what is natural to the soil. Man can raise the soil's stock of nitrogen or of any other nutrient, by artificial means: as it were, upon a platform; but he must then be prepared to hold up that platform by his own efforts, against the natural forces which are always tending to pull it down. The situation of agriculture is precisely analogous to man's holding up a real platform or sheet of matter – the soil's collective agencies taking the place of gravitational force acting upon a material object artificially raised above its natural level. [Fig. 5, p. 209.]

To return to the phosphate, of which there normally is a small stock in the soil minerals. Phosphate can be released from the soil minerals or rock phosphate only slowly. If phosphate is to be available to plants in excess of the natural supply from the minerals to the soil solution, it must be in the form of phosphate recently added in compounds which can be attacked by the soil microbes (as organic matter of plant and animal tissues, including bones) or else in some other form soluble with comparative readiness in the soil solution, and possibly without participation of the soil microbes in making it soluble and available.

Even in good, open, well-aerated soil, the soil and the soil solution consist of a distinctly odd-sounding jumble of what is soluble and what is not soluble in a solution of calcium bicarbonate. There is calcium sulphate in fair quantity in the solution, and some excess carbon dioxide, and a very little oxygen and nitrogen as gases. There may be a little nitrate – depending on the ability of some bacteria to produce it, and of other bacteria to destroy it; as well as on the ability of rain and plant roots to remove it. There will be small amounts of potassium and other metallic ions in the solution; more ions will be attached to the colloids, and available to pass into the solution if and when conditions permit. Practically all the ammonia and most of the phosphate will not be in solution at all (they will, however, be potentially available to the extent that the ammonia is attached to the colloids and/or is being released from organic matter by bacteria, and to the extent that the phosphate is in recently-added organic matter or exists as a recently-formed precipitate). There will be (let us hope) small lumps of calcium carbonate and sulphate waiting, as it were, to be dissolved; and the rest of the minerals (pretty tough ones, mostly) being slowly attacked by the soil solution.

There will also be some reserve of all or most of the nutrients in organic matter not yet decomposed. The whole system of solids + solution will be bathed in an atmosphere very much like that of the above-ground air, but containing a higher and much more variable percentage of carbon dioxide, and saturated with moisture.

CHAPTER ELEVEN

How shall we Feed the Plants?

> 'A barrowful will do, to begin with.'
> 'A barrowful of *what?*' thought Alice.
> *Alice's Adventures in Wonderland*

THE non-leguminous crops are going to supply a little of the protein and most of the carbohydrate required by animals. The clovers, lucerne, beans, and other leguminous crops are going to supply most of the protein – and virtually all of the bulk protein – together with some of the carbohydrate, which animals need. In addition, and by virtue of their unique ability, among crop plants, to gather nitrogen from the air, the leguminous plants well cultivated will add nitrogenous residues directly and indirectly to the soil to an extent beyond that natural to the soil, and will thus not only meet, from the soil and from the air, their own needs in nitrogen, but will furnish nitrogen and soil organic matter for the non-leguminous crops such as wheat, oats, barley, maize, turnips, sugar-beet, and others which come after the legumes in a rotation of crops; and the legumes will also supply nitrogen to the grasses and useful weeds grown alongside clovers or other legumes of grassland.

For all this increased growth of animals and plants – increased, that is, to a level well above the natural – the problem is to supply the earthy nutrients required by the crops of the whole rotation, and of the legumes especially; and – if still more carbohydrate is desired – to supplement the nitrogen which is made available by leguminous plants for their non-leguminous companions and successors.

It should not be necessary to repeat either the inference that we have reached a quite artificial situation, or the fact that cultivation of leguminous crops is the crux of intensive farming and of 'high-pressure' supply of food for man. But I do repeat them, just in case. . . .

I may put it another way. Because of the unique ability of leguminous crop plants to convert nitrogen – from the soil nitrate or ammonia, or from the air – into bulk protein for feeding animals and other purposes above-ground, and for addition to the stock of organic matter and of plant-available nitrogen in the soil, the crux of food-supply on the modern scale is to render enough of the earthy nutrients, and of lime, phosphate, and potassium especially, available to the legumes. If the supply of plant-available earthy nutrients is assured, then (assuming sufficient water and good aeration in the soil) the supply of protein will look after itself in proportion to the earthy nutrients of plants.

This matter of *availability* of nutrients is of the highest importance in all spheres of feeding. For animals it is largely, though not entirely, a question of digestibility of what is provided. It is obviously of no use to feed animals on leather or other indigestible material, no matter what proportion of total nitrogen or total protein may be recorded by a chemical analysis. It is a matter of common sense in animal rationing to have regard to what is digestible, and not to the total, of a nutrient or mineral element. I feel that most people will readily agree to that common-sense proposition, which is but one facet of the truth that you cannot feed animals with arithmetic. Yet it is surprising how often this simple and all-important matter of what is available is lost from sight, in favour of the chemical total.

Why is it, I wonder, that few of the people who freely indulge in controversies over the nutritional value of various forms of flour and bread ever have regard to anything but the gross percentages of the nutrient – be it calcium, phosphate, iron, or vitamins – they select for discussion? For various reasons the proportion available or utilizable or net may be only a small positive fraction of the total; sometimes (though not, I think, in wheat flour) the presence of a relatively high percentage of, say, calcium in the foodstuff may set up what is called a negative balance: *i.e.*, eating the food may set up a demand for extra calcium to be added if the eater is not to be despoiled of calcium already obtained from elsewhere. Rhubarb is such a food; and there is sound reason for taking milk or custard with it – to supply additional calcium to satisfy the demand set up by ingestion of rhubarb; and, of course, there is good reason for taking more milk than that minimum, to the end that one may get a positive benefit from the calcium of the milk (or, from the calcium of the milk plus rhubarb, if you prefer it put that way).

Controversy about such matters as the amounts of nutrients in various grades of flour seems to be unusually vapid and to lose sight of the common sense about availability, except on the rare occasions when the lists are entered by people able to read at least some part of the *Biochemical Journal*. Yet, I do not really think that a high standard of scientific training is necessary for one to grasp the point that only the nutrients that can be taken up and assimilated by an organism are of any use to it. The rest might as well be on the moon; so to argue about totals is useless; unless one knows and understands that the total nutrient in question is indeed in the rather uncommon condition of being wholly available.

We can dismiss the question of the availability of the minor nutrients in flours and breads as being, on the whole, futile; since (for one thing) there are means of supplementing them by eating other things besides bread and pastry. The question of the availability of plant nutrients which may be in the soil, or can be added to it, is immensely important. The existence of many of us depends upon it.

The words *available* and *availability* are ambiguous, and may be taken to mean what is there: the total, or what is in a deposit. Confusion between what is commercially available and what is available as a nutrient for plants has not infrequently darkened counsel. In the known deposits of potassium salts and of rock phosphate there is enough potassium and phosphate to meet the present-day requirements of the world for a thousand years; and though it is unfortunate that the British Commonwealth possesses no really large and accessible deposit of either, it can be said that in spite of political divisions the commercial availability of potassium salts and rock phosphate is good throughout the technically-civilized world.

Plant-availabilities of potassium and phosphate, however, make very different and contrasted stories. That is true, both of what is naturally present in soils, and of what can be added to soils by way of fertilizer so as to obtain that extra sustentation of protein and other foods which we either hope for, or take for granted, according to the limitations of our knowledge and awareness.

I have several times used the term 'plant-available' to mean 'what is readily assimilable by plants'. The point is that the total amount of, say, phosphate or potassium in a soil is always many times the amount that a crop grown on it can assimilate during one season. There is likely to be a big enough total of each of the principal earthy nutrients to supply the needs of several hundred annual crops; but, as I have written on page 61, most of it may not become available for centuries. There is usually enough total nitrogen in the soil's organic matter for many crops; the period during which it may be available (though likely in diminishing amount annually) will probably not be reckonable in centuries, but rather in decades; still, that will not make a profound difference, since the nitrogen supply in and to soil is renewable to a limited extent through growth of nodulated leguminous plants and free-living nitrogen-fixing bacteria even in the wild; so that under natural conditions a balance will be struck between vegetation-cover and actual annual availability of water and all nutrients.

That natural balance has become inadequate to meet the demands of civilized peoples. It is therefore important to look closely into what has been done to augment the supplies of plant-available nutrients in the soil: to ask why we are here, where we have got to, and what, if anything, can be done about it now we are where we are.

CHAPTER TWELVE

'Inexhaustible Nitrogen of the Air'

'It was a glorious victory, wasn't it?' said the White Knight, as he came up panting.

'I don't know,' Alice said doubtfully. – *Through the Looking-Glass*

LET us start with nitrogen: the favourite of politicians, to whom it offers a facile remedy; of political economists, who are quite good at arithmetic; and of writers about 'man's conquest of his world' or 'harnessing the elements': to all of whom (politicians, economists, writers) what they know about it provides a grand story. They usually don't know anything about nitrogen except synthetic nitrogen captured, by methods discovered during the present century, from the air over factories. The word 'clover' is not in their vocabulary (it seems). Admittedly, clover is an unimpressive plant. The Second Agricultural Revolution, which happened this century, and which has stocked the agricultural world with more protein than was ever in it before, was so quiet that it hasn't been noticed. Nitrogen in the forms of ammonia and nitrate is wholly available to every kind of plant. That is true whether the compounds are released or made by microbial action in the soil from the soil organic matter or from organic matter such as farmyard manure or urine or poultry manure added by any means to soil, or whether it is added as fertilizer of natural fossil origin (such as Chilean nitrates) or made artificially by distillation of coal or otherwise. If the nitrate – from any source – is not washed out it can immediately be taken up by plants. If ammonia or ammonium salts are present or supplied, they will either be taken up as such by plants, or be converted (under favourable conditions) to nitrate by the nitrifying bacteria. If any other source of nitrogen is given or is already in the soil, it will be more or less rapidly attacked by miscellaneous bacteria, with the result that ammonia is set free for uptake by plants, bacterial conversion to nitrate, absorption on the soil colloids, or (if a great excess is present) loss by washing out or other means.

Thus, apart from losses and removal, all common compounds of nitrogen normally in good soil or added to it may be said to have the whole of their nitrogen available to all kinds of plants. It does not matter to any plant whence it obtains its nitrogen. Whether that nitrogen is in recently-dead matter or is of fossil or synthetic origin does not make any difference to its ultimate availability to any kind of plant; but of course to give nitrogen in ammoniacal or nitrate form in the form of soluble fertilizers is the quickest method of putting a large amount of available nitrogen at the disposal of plants, and so of hastening their growth if other factors are not limiting.

Unfortunately, we have to accept the fact that even quickly-available nitrogen does not form protein in bulk except in leguminous plants of all ages and in the young and very recently-grown parts of other plants. It is rather costly and would be wasteful to give much commercial nitrogenous fertilizer to leguminous crops, except for crops of special nature and high monetary value, such as early green peas for market. Nor, except for such special purposes, is it often done to give much nitrogenous fertilizer to legumes grown alone. A mixture of fertilizers containing a high percentage of nitrogen is often applied to grassland (the amount of 'potato manure' used in Scotland during the war – when it was forbidden to treat grassland generously – was said to be much greater than was needed for the whole acreage of potatoes: I wonder where it went?); but for 'pure' leguminous crops such as field peas and soya beans it is usual to use a mixture containing a low percentage of nitrogen, or none. You will legitimately infer that it is usual to supply all leguminous crops – including grassland – with important quantities of plant-available phosphate and potassium.

That a characteristic function of nitrogenous manures is to yield non-nitrogenous constituents of plants (*i.e.*, carbohydrate and oil) is a statement published by the ever-percipient J. B. Lawes in 1900, and was probably known to him much earlier. As a matter of world-wide practice, use of soluble nitrogenous fertilizers on a large scale is general for starchy and sugary crops, also fibre crops, and non-leguminous oil crops. Lawes's finding holds good for every kind of nitrogenous nutrition of plants. When ammonia, ammonium salt, or a nitrate is supplied to non-leguminous plants, whether by the operation of the microbial successions of soil or by any other means, it does not produce much protein. Some protein is necessarily produced as an accompaniment to the new and vigorous growth, and as part of the tissue which will envelop the starch grains and oil cells when the plant has matured and stored its carbohydrates and fats. In the mature non-leguminous plant that extra protein is, as it were, diluted with a much greater increase of stored carbohydrate and fat.

Especially in recent years, great hopes have been unsubstantially built upon the recently-acquired ability of man to fix nitrogen from the air in factories. In building hopes for perennial production of food 'from the inexhaustible nitrogen of the air', whose artificial utilization has been held to make the world independent of wasting assets such as Chilean nitrate, several weighty factors have been lost from sight.

One has been mentioned; still, carbohydrates and vegetable oils can be foods. But to talk about synthetic nitrogen-fixation having brought about an increased output of food, or being capable of still further development for feeding the world's hungry millions (or any such phrase), is to overlook the necessity of balance between carbohydrate and protein. Given yet more starch and sugar from more intense manufacture and use of nitrogenous fertilizer – synthetic or other – will the problem of feeding the world be solved without other equally far-reaching and possibly more complicated adjustments to supply protein to match? Or, at the lowest level, will more nitrogenous fertilizer effectively produce greater yields in most regions unless more lime, potassium and plant-available phosphate can be given to the soils to sustain the extra growth of plants induced by the extra nitrogen supplied to them?

The answer to both these questions is, No. Artificial nitrogen-fixation by itself can solve no general food problem. Reliance upon synthetic nitrogen to the exclusion of other important ingredients of the problem of increasing crop growth could only mean that the other components would make themselves felt insistently.

It takes a chemical to *make* a chemical. Energy alone can make nothing material of importance in everyday affairs. The oldest successful process of synthetic nitrogen-fixation is the Birkeland-Eyde, used since 1905 in Norway in a district of very abundant water-power. There, hydro-electricity has been put to make a huge arc, over which a current of air is passed, with the result that the torrefied air partly combines to make oxides of nitrogen, and those, passed into water along with more air, form nitric acid. Thus it is possible to make a compound by using only the renewable materials air and water; but nitric acid is practically the only compound which can be made by application of electrical energy to renewable materials.

Nitric acid is not of much use by itself; and to convert it into a transportable fertilizer it is necessary to combine it with something else obtained from the earth, such as chalk. Almost every other synthetic nitrogen-fixing process except the Birkeland-Eyde depends on use of coal, both as source of energy and for chemical reasons.

In the whoopee of returning thanks to the ingenuity of chemists and technologists who, in perfecting these processes of nitrogen-fixation, have freed the world for ever from fears about the eventual exhaustion of fossil nitrate deposits, it seems to have been quite forgotten that the consumption of wasting assets has merely been shifted: from nitrate to coal, in general, though sometimes natural gas can be used instead.

Since about five tons of coal are needed to fix one ton of air-nitrogen – or not less than three-quarters of a ton of coal per ton of ordinary nitrogenous fertilizer – the best that can be said about economy (if that is the word) of wasting assets is that coal is better able to bear the rate of assault now prevailing than is Chilean nitrate. Much more combined nitrogen is now used annually throughout the world than was consumed in the days before synthetic nitrogen-fixation was evolved; and much the greater part of the world's present output of simple nitrogenous compounds like ammonium salts and nitrates has been made artificially – by burning coal. Incidentally, and incredible as it may seem, a greater tonnage of mined Chilean nitrates is now used annually for agriculture (as fertilizers) than was so consumed about fifty years ago.

Synthetic fixation of nitrogen on a large scale in factories has been widely acclaimed as one of the greatest feats of this century. I do not propose to dispute the greatness, or the novelty, of the technical achievement. What I wish to do is to put its implications into proportion as regards food production.

Synthetic nitrogen-fixation has been of the greatest value to belligerents not able to get enough, or any, ready-made simple nitrogenous compounds from Chilean deposits or other natural sources, or from distillation of coal. It has made obsolete the use of fossil nitrate in manufacture of nitric acid for explosives and all 'heavy' industrial purposes, including the manufacture of that very important chemical, sulphuric acid. It has enabled us to increase appreciably the output of sugar from cane and beet, and of potatoes, rice, maize, cotton, and other crops which accumulate sugar and farina accompanied by very little protein; also of tea. The contribution of nitrogenous fertilizers of synthetic and fossil origins to food production has been and still is considerable; yet, nutritionally, it will be seen that our new-found ability to fix air-nitrogen synthetically to any extent compatible with our reserves of carbonaceous fuels is capable of disappointing those who do not know of Lawes's generalization or of the actual practices of efficient agriculture.

Misplaced enthusiasm about the influence of synthetic nitrogen upon food-production is mostly shown, I think, by economists and sociologists; also by chemists unconnected with agriculture. The best-known example of a chemist basing exaggerated hopes on synthetic nitrogen-fixation is William Crookes. The echoes of the part of his address in 1898 to the British Association that dealt with the prospects for increasing the supply of wheat through recourse to synthetically-made nitrogenous fertilizers have not died down yet; they are still sometimes caught up, distorted, and amplified by people who have not troubled to read what Crookes did say, or to compare his fancies with actualities. As Stephen Leacock remarked: In a controversy, a half-truth, like a half-brick, is better than a whole one: it travels further.

Crookes made a reference to bacterial nitrification, though he did not follow up the implications. At every other material point of his excursion into agricultural science Crookes was off the rails.

He looked forward to an immense increase of wheat-production if the 'new' countries had new supplies of nitrogen put at their disposal by artificial fixation. Some people think that he has been justified by events. In fact, the great increase of output of wheat that has occurred since about 1900 has had nothing to do with artificial nitrogen-fixation; it has been achieved from opening-up of the semi-arid lands which were almost untapped when Crookes spoke.

On the former prairies of North America now cropped with wheat, no fertilizer of any kind is used, or could be used with advantage (except in conjunction with irrigation: page 161). In Australia, substantially the only fertilizer used for wheat or other cereal on the semi-arid lands is phosphate, as superphosphate.

The point about nitrification is very important. Whenever the nitrifying bacteria of soil convert ammonia or ammonium salts into nitric acid in the ordinary and normal course of their bacterial business, a demand for calcium or other metallic element able to neutralize that acid is set up.

If the demand for calcium or other metallic base[1] is not met, the soil becomes acid; this impedes both the growth of crops and the activities of the majority of kinds of soil bacteria. This tendency towards acidification through bacterial production of nitric acid from ammonia in the soil is a normal and inescapable phenomenon of good soil. It operates whatever the source of the ammonia – whether from decomposition of organic matter or anything else, or from the action of man in supplying ammonia or ammonium salts as such to the soil. It is not a stigma peculiar to fertilizers of any one origin. It operates equally with poultry manure – which is essentially an ammonium salt – as with ammonia derived from coal or any other source.

An apparently paradoxical result is therefore obtained when ammonia, or a solution of it, is added to soil. A solution of ammonia is alkaline; yet, owing to the onset of nitrification, ammonia applied to soil – having been converted by bacteria to its chemical equivalent of nitric acid – has the same effect on soil as if the soil had been watered with the corresponding amount of free nitric acid.

The recent use of liquid anhydrous ammonia as fertilizer in the United States has been so slick and spectacular that knowledge of it is becoming fairly widespread in Britain. Some people whose ideas about the distinctions between technology and science are not very clear are beginning to think that British farmers are behindhand in not trying to use liquid ammonia; and that once again we in Britain are lagging in the application of agricultural 'science'. I fancy that British agricultural science is second to none.

There are several reasons for the increasing popularity of synthetic liquid ammonia in the U.S. Some of them recall the 'easy-wayers' rather than those who take the virtuous path.

1. *Metallic*, because ammonia, though chemically a base, will produce acid and will also lead to further loss of calcium by operation of base exchange.

Though ammonia applied to soil can produce as great a decalcification as six times its weight of ordinary concentrated nitric acid, U.S. practice does not seem to have fully taken that fact into account: trusting, perhaps, that the cheque for metallic bases that is being drawn will never be presented.

Some of the factors are commercial. Not all fertilizer manufacturers in the U.S. are well situated to supply metallic bases; another factor is the extra cost of transport when lime and other earthy materials are bought to balance the nitrogen of ammonia. On the whole, prudence suggests that one should pause before becoming an enthusiastic advocate of Britain's copying this U.S. development just because it is American or seems to be the latest thing or because wonderful reports about it have been received.

Chapter Thirteen supplies some more ammunition. If nature receives her dues in the United States, an annual consumption of nearly 100,000,000 (one hundred million) tons of limestone, or its equivalent, may be expected there before very long: historically speaking.

Whenever ammonia or an ammonium salt is applied to soil, or is formed in soil, there is a loss of metallic bases additional to what is brought about by nitrification-acidification. As already mentioned, ammonium is held by the soil colloids. It is not then easily washed out – until it is nitrified; though the soil colloid is probably the main seat of nitrification. Ammonium, or actual ammonia, gets from the soil solution on to the soil colloids by turning out some calcium or sodium or other 'base' or kation from its loosely-established position on the colloid surface. The metallic base is thus set free into the soil solution, and is very apt to be washed away.

There are thus two sources of loss of base (able to neutralize acid) ascribable to nitrogenous matter in soil. Base-exchange operates when ammonia or an ammonium salt is present or is formed naturally, or is added to soil; nitrification operates upon every nitrogenous compound – except nitric acid or nitrates. Formation or addition of ammonia, or its compounds, leads to the onset and progress of both these causes of acidification or decalcification of soil. Consequently and inevitably, a requirement for additional metallic base in the form of some earthy substance is set up, if the soil is not to become progressively acid.

We are therefore confronted with the situation that every form of nitrogenous compound added to soil by any means will make the soil poorer in bases, and actually or potentially more acid, unless the nitrogen which is given is accompanied by its equivalent of metallic base: *i.e.*, *earthy* material able to neutralize acid. In brief: the more nitrogenous fertilizer, the greater the demand for alkaline earths. To this coupling there is no exception.

The effect is slightly mitigated by the circumstance that plants take up more nitrogen, to form their proteins, than what is equivalent to the bases assimilated by the plants. Nevertheless, we have in theory (and in practice on uncropped humid soil) the circumstance which will seem very remarkable, and perhaps incredible, to many chemists: namely, that one equivalent of nitrogen as ammonia can displace from soil two equivalents of base. One atom of ammonia-nitrogen is able to turn out one atom of calcium or magnesium! If this fact had been presented to chemists a hundred years ago, there would probably have been some invocation of 'vital activity' or other occult force to account for the apparent departure from the fundamental laws of chemistry. There *is* participation of living organisms – the ammonifying and nitrifying bacteria; but there is no need to look for mystery. The facts now known amply account for the observations.

In the estimates given above and later, and referring to losses of base and amounts of lime and earthy material needed to make good those losses, both the causes of acidification mentioned on page 142 have been taken into account.

The practical point arises from the fact that in introducing synthetic nitrogen-fixation we have not escaped the operations of the laws of nature. There is no way of getting round them! In this particular instance, by learning to fix 'the inexhaustible nitrogen of the air' synthetically, or by calling upon other sources of nitrogen to make ammonia or nitric acid for fertilizers, we have not extricated ourselves from dependence upon earthy materials such as chalk and soda. That remains true even if coal is not used as well.

Chemically and microbiologically, it does not matter in the least how the earthy base is supplied to offset acidification from nitrification and base exchange in the soil. The commonest and cheapest base is lime. That may be naturally present in the soil (as carbonate) or in irrigation water (as bicarbonate); and if the soil or water is highly calcareous there will be no need to worry about acidification – yet.

Lime may be supplied by the farmer as part of his ordinary liming programme; it may be mixed or combined with the fertilizer at a factory; and the necessary base may have any other origin – such as the sodium in Chilean nitrate, or dust blown from a volcano or a lime-works; but metallic base *must* be supplied if the soil is not to become more acid through use of nitrogenous fertilizer or manure of any kind.

It may be objected that farmyard manure (for instance) does not make the soil more acid; whereas it is common knowledge that 'artificial' fertilizers have often done so. The answer is that it can be only misleading to try to create an antithesis between 'natural' manures and 'artificial' fertilizers. Farmyard manure happens to contain enough base to satisfy the acid produced during the nitrification of the nitrogen in the manure; ordinary urine, or poultry manure, does not. Acidification of soil from use of ammonium fertilizers has occurred to a quite notorious extent on non-calcareous soils through failures of farmers to apply enough lime to neutralize the acidity created. Poultry husbandry experts recognize that the soil of a hen-run or a field on which many poultry graze becomes acid, unless a relatively heavy dose of lime is spread. So must lime be given when 'liquid manure' (= urine) is applied.

Any programme for building more factories to make more nitrogenous fertilizer from the air or otherwise, in the hope that food-production from non-calcareous soils will thereby be increased, must be matched by correspondingly strenuous provision for an equally generous supply of lime or other earthy base.

This is a consequence often overlooked. If earthy matter enough to match the extra nitrogen is not given, soils will become more acid to the extent of the deficiency; and soils which are not naturally calcareous will fall away in crop-producing ability. The rate of this decline in productivity will be proportional to the size of the gap, or divergence, between supply of extra nitrogen and supply of earthy material to correspond.

This argument applies only to the earthy material required for neutralizing soil acidity. It takes no account of those other, different, but not less imperious, demands for earthy materials such as potassium and phosphate. More of these earthy nutrients of plants will be needed to sustain additional growth of crops for feeding extra numbers of people. There will anyhow be a continuing demand for potassium and phosphate and their earthy concomitants (including fuels), merely to keep alive as many people as there now are.

CHAPTER THIRTEEN

The Rabbit sends in a Little Bill

'How am I to get in?' asked Alice again, in a louder tone.
'*Are* you to get in at all?' said the Footman. 'That's the first question, you know.' – *Alice's Adventures in Wonderland*

SUPPOSE a Plan were to be promulgated for increasing the output of food by building more factories for fixing nitrogen from the air. A grand Plan for a total of a nice figure like a million tons of nitrogen to be brought down annually from the inexhaustible nitrogen of the air in five or six super-factories each costing eight or ten million pounds and 'backed by all the resources of science' (*sic*) would be attractive, and would lead the public to think that something was really being done to defeat the challenge of famine, exhibit man's mastery over nature, and so on, in phrases I need not further anticipate.

I am afraid the Plan might stop there. Whether it provided for the earthy concomitants or not, a million tons of nitrogen used annually as fertilizer would entail very considerable extras to current consumption, if the extra nitrogen were used in conjunction with a fair state of mixed husbandry, and if the soil was not to get worse instead of better.

For example: it may be assumed that a six-course rotation is in use, that livestock are already kept (so that the first course can have a fair dressing of dung); and that some fertilizers are already used, so that the additional fertilizers will bring additional production needed to make good use of the resources of water and soil. It is also assumed that the Plan provides for half of the new synthetically-processed nitrogen to be in the form of a mixture of roughly equal parts of ammonium nitrate and chalk, the other half being ammonium sulphate: this being a condition which is favourable to manufacture of synthetic nitrogenous fertiliers, and brings in the possibility of dispensing with a call upon the visibly small reserves of mined sulphur or pyrites for production of those kinds of nitrogenous fertilizer.

The rotation and its extra fertilizers – including the new nitrogen and no more than the earthy materials which its use would entail – would be somewhat as set out below. The fertilizers are stated in units of hundredweights (112 lb.) per acre. It should be noted that the table is not intended to present a system or detailed recommendations for a programme; thus, the only potassium fertilizer mentioned is the chloride, and it is not meant that this is ideal for all purposes or that it would in practice be used. Nor does the omission to mention basic slag imply anything except that the table has been presented in simple form for the purpose of facilitating calculation, and putting the bare structure of the argument before a non-specialized reader.

Year	Crop	Fertilizer		
		Nitrogenous (half of each as above)	Potassium chloride	Super-phosphate
I	roots, dunged (potatoes, sugar-beet, turnips, kale)	4	2	3
II	cereal (oats, maize, rye, barley)	2	–	1
III	grass } during			
IV	,, } three years	2½	2½	5
V	,, }			
VI	cereal (wheat, barley)	1½	½	–
	Total	10	5	9
	Per annum, average	1½	0·8	1½

Notes. It is not important whether the tons are short (2,000 lb.), long (2,240 lb.) or metric (2,204 lb.). If metric tons are assumed, the figures for the hundredweights can be taken as half-quintals (50 kg or 110 lb.); or roughly as quintals per hectare. 1 hectare = 2·4 acres.

A million tons of nitrogen in the forms assumed would

make roughly three million tons of ammonium nitrate plus chalk, and two and a half million tons of ammonium sulphate: a total of five and a half million tons of nitrogenous fertilizer, which at the rate of about 1½ cwt per acre per annum would annually dress 70 million acres, or about 110,000 square miles. This is a little more than half the total area of continental France and Corsica. The extra production gained by use of the fertilizers could be expected to feed perhaps fifty or sixty million people. That this rough estimate is not unreasonable can be judged from such facts as that France is *about* in equilibrium of forty millions, having no very substantial exports or imports of food; most of the area of France is cultivated, the greater part being either upland or under peasant cultivation at not far above a good subsistence level; only a small fraction of France is under intensive cultivation with an already high consumption of fertilizers.

The bill of fertilizers wholly or practically necessitated by the extra million tons of air-nitrogen would be, in millions of tons per annum:

	Nitrogenous	Potassium chloride	Superphosphate	Chalk
Tons of fertilizer	5½	2½	4½	6

This is a total of 18½ million tons needed to satisfy the execution of a programme of utilizing one million tons of nitrogen synthetically from the air. About a quarter of the chalk would be already mixed with one of the nitrogenous fertilizers. The carbon of this (synthetic) chalk would represent part of the coal used in the synthetic preparation of nitrogen from air by the usual Haber process.

The bill for elements, and for some of the raw materials consumed would be, in millions of tons per annum:

	Nitrogen	Potassium	In superphosphate Sulphur	In superphosphate Phosphorus
	1	1·2	0·4	0·8
And coal (carbon)	4			
(as sulphuric acid)			1·2	
(as phosphate rock)				2·4

If the ammonium sulphate were to be made with use of sulphur (instead of gypsum as was assumed) a further 0·6 million tons of sulphur would be needed – making 1·0 million tons of sulphur per annum in all for the programme. This would require a total of 3 million tons of extra sulphuric acid annually (*i.e.*, beyond what is now made). It will be seen that to burn elemental sulphur in conjunction with manufacture of ammoniacal fertilizers – whether synthetic or a derived by-product of the distillation of coal in gas-works or coke-ovens – is relatively more costly than to burn sulphur for the purpose of making a phosphatic fertilizer.

Mined sulphur is not indispensable. Phosphatic fertilizers *could*, with difficulty, be made without sulphur; or elemental sulphur could be obtained from gypsum. In either case it would be necessary to consume an extra mass of coal about equal to the weight of sulphur produced or replaced.

No coal or other debit of fuel has been charged against preparation of potassium fertilizer, which is relatively economical in fuel and power; nor has any fuel or power been charged against grinding the limestone or chalk.

These calculations survey the additional fertilizer and chalk or limestone and coal and sulphur that would be required each year if an additional million tons of nitrogen were to be fixed synthetically per annum, given a fair balance of fertilizer applied to soil on a reasonable rotation, followed in a humid region on average land neither very poor nor notably rich in any nutrient. This is the sort of land now farmed in many parts of Europe with some fertilizer, and capable of giving better yields and of carrying more animals if it were better treated – and better managed.

The better treatment envisaged would *not* be simply the supply of additional fertilizer. The extra fertilizer would make it possible to grow higher-yielding varieties of crops and to carry additional livestock, making it possible to raise production all round (if the farmers had the extra capital and could meet other requirements beyond the scope of this book to discuss).

The main points are (1) that to turn a greater tonnage of nitrogen into fertilizer would set up far-reaching new responsibilities; and (2) extra fertilizers will not be sufficient to increase production of food unless a system of farming to correspond is adopted – otherwise the extra fertilizer would largely be wasted: at best in grass not grazed or in straw for which no market or use could be found; at worst in burial or unnecessary wastage in the soil.

The calculations may be taken as applying to land now under cultivation, or to new land starting from scratch after reclamation. The rotation is a sort of average, and will not fit all circumstances. The 'grass' is, of course, meant to be grassland bearing a mixture of grasses and clover and possibly other legumes, and a sprinkling of herbs to contribute variety and a certain mineral content beyond that carried by the graminaceous plants. If lucerne were to be taken instead of grass, less nitrogen could be given – the nitrogenous fertilizer could be spread more thinly over a larger acreage – but more phosphate and potassium would probably be needed. On calcareous soils it might be possible to dispense with adding chalk.

It is important to note that the previous calculations provide for no more than the amount of chalk or limestone needed to satisfy the inexorable demands set up by base exchange and the activities of the nitrifying bacteria. To the 2 cwt of chalk per acre required annually to offset the acidifying effects of about $1\frac{1}{2}$ cwt of nitrogenous fertilizer supplying ammonia or nitrate gained from the air alone (or from distillation of coal), there would have to be added in most humid regions (at least where the soil is not calcareous) a further minimum of 2 cwt of chalk or limestone per acre per annum to offset the effects of ordinary leaching.

With that doubling of the requirement for chalk or 'lime', the figures suggested as additional consumption of fertilizer do in fact come close to what is now used on land farmed moderately well, though not highly or really intensively. The figures per acre present, in short, a rough average for present-day farming which approaches the intensive.

Totals of materials, in millions of tons, to be found for an additional one million tons of air-nitrogen per annum, would be something like this:

	(*a*)	(*b*)	(*c*)
Potassium fertilizers	2·2	2·2	2·2
Phosphate rock	2·4	2·4	2·4
Sulphur	0·4	0·4 0·6	
Coal (*d*)	4	4	4 3
Gypsum or anhydrite	2?		5?
Sulphuric acid 100% pure[1]	1·2	3·0	?
Chalk or limestone 100% pure[2]	12	12	12
Total material to be moved in manufacture for delivery of 12 million tons of fertilizers and 12 million tons of chalk to farmers	24	24½	28½

(*a*) Elemental sulphur burnt only for superphosphate.

(*b*) Elemental sulphur burnt for superphosphate and ammonium sulphate.

(*c*) All sulphur derived from, or replaced by, use of coal.

(*d*) A conservative figure; a coal-equivalent of 5 tons per ton of nitrogen would be truer.

(1) In which is included the sulphur quoted above. The tonnage of commercial acid would be rather larger.

(2) Including 1½ derived from coal during synthetic nitrogen-fixation.

Some of these materials might require to be moved thousands of miles (*e.g.*, sulphur and phosphate rock); others, like sulphuric acid, no further than from one part of a factory to another. Fuel used for ships or other transport, or represented by depreciation of steel, etc., is not included. A factory capable of fixing 100,000 tons of nitrogen per annum would need 100,000 tons of constructional steel.

We therefore have the rather surprising conclusion that on well-farmed land in the United Kingdom or western Europe, or other area which sustains a dense urban population at a fair standard of living, with or without imports of animal feeding-stuffs or human food – each acre is getting every year something like the following quantities of materials, besides dung at an average rate of about a couple of tons per acre:

	Cwt
Nitrogenous fertilizer	1½
Potassium fertilizer	¾
Phosphatic fertilizer (superphosphate or basic slag)	1½
Limestone or chalk	4
Total, about	8

One person living on a mixed diet including meat, milk, and eggs requires *about* 2½ acres of land annually to be under cultivation. No precise figure can be given; the acreage needed to support one person under decent European conditions may be as low as 1½. Under minimum subsistence conditions, a couple of acres or so can support a family; so an estimate of 2½ acres per person may not be far wrong for generous conditions of living.

It will be seen that this requires a ton of inorganic material to be applied to the land every year for each person. If production from the land is to be increased, the minimum that can be asked is that such rates of consumption of fertilizer, and, where necessary, of lime, shall be attained. Some nineteen-twentieths of the additional burden will have to be borne by materials dug out of the earth.

CHAPTER FOURTEEN

Introduction to Vulcan

'Thinking again?' the Duchess asked, with another dig of her sharp little chin.

'I've a right to think,' said Alice sharply, for she was beginning to feel a little worried. – *Alice's Adventures in Wonderland*

WE have seen in Chapter One that there is no point in 'feeding the (collective) soil bacteria'. The microscopic organisms of soil, rivers, and other natural habitats, constitute equilibrium-systems. To feed one of these systems, in soil, or in, say, a puddle, with a single substance such as sugar or ammonia, or a material of unspecified composition such as compost or other organic matter, is simply to alter the equilibrium to a corresponding, usually unknown, extent; and for a time which cannot be specified by us.

The only plant-nutritional purpose served by adding organic matter or any other chemically complex material to the soil is that the soil microbes shall ultimately release nutrients for plants, from the non-living material. That function of soil microbes – like the other purely decompositional and oxidative functions of the mixed microbial population of the soil – is not indispensable. It can be replaced or got round, or supplemented, by adding ammonia or nitrate or phosphate or sulphate out of a bag or other container filled in a factory. The main functions of organic matter added to soil are physical. Those physical effects are modes of improvement of soil structure by means which are beyond the scope of this book, though they have a largely microbial genesis.

The function of soil microbes for which no artificial substitute has yet been found is the reduction of air-nitrogen to form important quantities of protein edible directly by man or by his domestic cattle. That function is practically the exclusive property of the legume-nodule bacteria which are in association with their leguminous hosts.

The legume-nodule bacteria have one striking and extraordinary microbiological character. While in the nodule they exist as a strictly pure culture: each nodule contains exclusively one strain of one species. For that reason, consideration of the legume-nodule bacteria, after they have entered their host-plant, lies outside discussion of the balance and struggle of life in the soil. It is interesting that these bacteria within their plant nodules should collectively present the greatest example of a pure culture in nature, wherein a conditioned conflict between a variety of species and forms is the rule, and pure cultures of any kind of microbe are the exceptions.

These nodules and their contained bacteria represent a highly specialized kind of withdrawal from the general conflict in the soil outside; they are little empires of their own. So long as the nodules last, they have their own allegiances and laws and nutritional problems – quite distinct from those ruling outside. That is not to say that they exist in a different world; it is to suggest that if there is one class of bacteria about the feeding of which we should worry, it is the legume nodule bacteria.

The immediate nutrition of the bacteria in their nodules is undertaken by the host plant. Whether air diffuses into the nodule from the soil, or is supplied by the sap, or in both modes, it seems that the plants do supply everything else but nitrogen; there is reason to believe that the nodule has a mechanism for controlling the utilization of oxygen. About actual 'feeding of the bacteria' – the uncomprehended mechanisms of supply of carbohydrates or organic acids to give energy for the bacterial nitrogen-fixation, and to serve as chemical carriers of the nitrogen after it has been fixed – there is nothing we can do except to ensure the well-being and vigorous development of the host plant.

'Feeding the nodule bacteria' has a meaning. Carbohydrate supplied by the plant to a well-found nodule will be utilized wholly by the strain of bacteria within it. There will be no competition from the generality of soil microbes in the outer world; and there will be no problem of apportionment of bacterial food supplies or other problem of equilibrium between different kinds of soil microbes. The host-bacterial relationship in leguminous plants is thus of a totally different philosophical kind from the relationships between microbe and microbe, and microbe and plant, in the soil outside the roots. I hesitate to say that the leguminous plant/legume nodule bacterial relationship is the simpler; but it is certainly different – if only because interaction of microbial species does not exist inside the nodules. Within certain limits of suitability of season and so on, carbohydrate or other organic matter will be furnished by the host plant to the nodule; such materials will be used within the nodule to sustain nitrogen-fixation, and hence to advantage protein-production within the plant – and possibly elsewhere as well.

The extent of bacterial nitrogen-fixation, and consequently the extent to which leguminous plants will form protein over and above an amount corresponding to what combined nitrogen they can get out of the soil, must depend upon the encouragement which we, as cultivators of leguminous plants, can give to the legumes themselves by giving them enough nutrients, water enough, and suitable soil conditions. That is to say, the extent of nitrogen-fixation, and the tonnage of protein we can wring from the soil, will and must depend in the first instance upon the assiduity with which we furnish lime, phosphate, and potassium salts to meet the requirements of the leguminous plants for those substances. If the soil happens to be rich in any or all of those materials we need not add them.

Instances of soils naturally rich in all four of the requisites water, lime, phosphate, and potassium are, naturally, hard to find. Their simultaneous presences are incompatible with chemical and physical equilibria. The three chief mineral or inorganic requisites for growth of legumes must be added by man except in those rare locations where they are brought by natural means to the soil from outside. To find humid soils so satisfactory in both the more important nutrients potassium and phosphate, it is necessary to search the world. The island of Java and its small neighbour Madura; Hawaii; the neighbourhoods of Etna and Vesuvius; and a few other pin-points on the world map, are a nearly complete list. Many other volcanic regions (Sumatra, next to Java, is an example) get mostly the wrong kinds of natural manuring, or for other reasons do not benefit from periodical additions of inorganic plant nutrients from external sources.

'It is clear', said Dr René J. Dubos, 'that the myth of Demeter and Persephone symbolizes the death of vegetation at the end of autumn followed by a resurrection of dead nature in the spring. It implied some form of belief in the continuity of life, a faith in immortality. . . . Microbes contribute some of the essential links in the endless chain which binds together the whole range of living forms. They, too, like the runners in a race, hand on the lamp of life.'[1]

The Proserpine myth (of which a variant is associated with the Greek-named town of Enna in central Sicily) may further be taken to suggest that Mediterranean peoples recognized an association between high productivity of land, and fire, as represented by the local volcanoes. There was much coming and going between Greece and Sicily; and Syracuse – a Greek colony – is said to have been the first city in the world to have a population of more than a million. Like the volcanoes of Java and Hawaii, the volcanoes known to the classical Mediterranean peoples are exceptional in emitting basic substances. The soil on the flanks of the volcanoes was and is free-working and rich; so that parts of the lower slopes of Etna now carry the densest rural population in Europe.

The ancients no doubt obtained sulphur from volcanoes and *solfatare*, but the occurrence of small amounts of sulphur

1. In an address at a banquet of the Society of American Bacteriologists; printed under the title 'Microbiology in Fable and Art' in *Bacteriological Reviews*, 1952, **16**, pp. 145–51.

produced by chemical action of volcanic gases should not mislead us into thinking that sulphur is mostly of volcanic origin. The great sulphur deposits of Sicily are probably older than Etna; and, like the much bigger deposits in the Gulf of Texas, are not associated with igneous rocks. Large deposits of sulphur seem to be of sedimentary nature, having most likely been produced by bacterial reduction of the abundant sulphates of sea-water. With the principal exceptions of gypsum, sodium and potassium salts, and copper and aluminium, most of the materials used in actual agriculture on a modern farm are or could be of biological – often microbiological – origin: including even the iron of the ploughs and tractors.[1] Bacterial processes leading to production of sulphur are going on to-day in the depths of the Black Sea (page 65) and in its shallow coastal lagoons called *limans*.

There is no reason to suppose that the ancients applied sulphur for purposes of agriculture in the field, though they may have used it for accessory purposes such as bleaching of woven straw.

The present linkage between Pluto and Ceres, or rather between Vulcan and whatever Fabian goddess of legumes there may be, is new in history, and has not had time to become appreciated for what it portends. Do you think of F.A.O. as an active volcano, preparing for an active distribution of millions of tons of fertilizing materials; or as a quiescent one, ready to decorate itself with a plume of smoke, and to speak if need be? Or perhaps you do not think of it, or any other proposition about food, as having any relation to Vulcan and his furnaces and workshops, and ours.

The connexion between food and manufactures is a long chain of the type of story about the house that Jack built. All plants require carbon dioxide from the air, and materials such as sulphate, phosphate, nitrate, and ammonia

1. Compare Hugh Miller (*The Old Red Sandstone*, 1841): 'How strange, if the steel axe of the woodman should have once formed part of an ancient forest!'

– whether those substances have been through the soil-microbial mill or not. They also require other elements of earthy materials, such as potassium, calcium, magnesium. *In addition*, leguminous crops can, but do not have to, make use of air-nitrogen. (Botanical text-books often go wrong here, by implication, when presenting the subject of nitrogen-fixation by nodule bacteria. That subject being treated independently of other aspects of plant nutrition, a student is left with the impression that such bacterially fixed nitrogen is the sole source of nitrogenous nutrition for nodulated legumes). All leguminous plants take up nitrate, and probably ammonium, from the soil in precisely the way ordinary plants do. If the soil's stock of nitrate and ammonia is ample enough, the leguminous plants will take it up preferentially; and correspondingly less air-nitrogen will be fixed: the nodule bacteria appear to share our disinclination for doing more work than is compelled by the circumstances.

To encourage leguminous plants to grow well it is necessary to have in the soil an ample supply of plant-available lime or calcium, potassium, sulphate; also phosphate. These are all earthy materials. Thus it is true of nitrogen-fixation, and of protein-production in particular, that it depends in practice *not* on the illimitable nitrogen of the almost ubiquitous air, but on the abundance and availability of certain earth-derived substances in the soil.

There is no great difficulty in supplying enough lime where it is needed; that is, mostly, in the cool humid regions with climates like those of Britain, north-western Europe, and eastern North America. Nor does potassium present any serious problem of manufacture or supply; nor would sulphate, if the sulphur were to be regarded simply as a plant nutrient: if it were to be given as ground gypsum, for instance. Of all these substances, and of raw rock phosphate, there is in fairly accessible places enough to meet present-day requirements for plant nutrition during a thousand years or so.

In respect of most crops in most situations, the trouble

about phosphate is that the phosphate of raw rock phosphate, during the millions of years it has been under the sea and under the earth, has got itself locked up into a form from which plants cannot easily detach the phosphate. So long as phosphate circulates and keeps moving in the plant-(animal)-soil cycle, the phosphate radicle is easily split off by the soil microbes from whatever organic matter it was attached to, and then, under conditions of good husbandry, a good proportion of it can easily be taken up by leguminous or other plants. Ground or processed bones are excellent fertilizers; but reliance on bones raises difficulties of collection, besides setting a limit to production of food – even if no leakage or wastage occurred.

CHAPTER FIFTEEN

Mostly about Brimstone and Burning

'I know *something* interesting is sure to happen,' she said to herself, 'whenever I eat or drink anything.' – *Alice's Adventures in Wonderland*

MAN must indeed be a little lower than the angels if his future depends on the supply of brimstone. A little thought given to our present situation will reveal that food-production has come to be very tightly bound up with industry and with consumption of fuels. This can be seen to be a matter of fact: without knowing anything about the central position of the leguminous crops, or about the necessity of meeting their requirements for phosphate and potassium and other earthy materials. With equal certainty it can be taken as true and apparently inescapable (in the present state of knowledge) that any extension or development of food-production by conventional or other techniques will be no less dependent on devotion of further chemical, engineering, and industrial effort towards additional consumption of fuels and other wasting assets.

Analogies with the past will not help us if they are no more than superficial. Additional carbohydrate has been won at various times – is still being won – without use of fuel or fertilizers. Ancient Rome brought bread-wheat from Africa. Historically speaking, the semi-arid 'new' lands – which produce, and can produce, cereals almost exclusively – were opened up not long ago.

It is pertinent to note in connexion with the theme of the historical rapidity of the often unremarked changes on the food front, that the two Canadian Provinces which now export the most wheat were before 1905 so sparsely settled that they were only Territories. Though Winnipeg continues as the great centre for marketing of Canadian wheat, and Manitoba last century had given its name to a world-famous grade of wheat, the centre of Canadian wheat-production has within the last thirty years shifted westwards beyond Manitoba into Saskatchewan and Alberta. Not so very long ago (before Winnipeg was of world consequence) the centres of gravity of Canadian wheat-production lay in Ontario and Provinces still further east.

Oklahoma was not admitted a State until 1907. While Manitoba, the Dakotas, Oklahoma, and the rest of the North American prairie wheat belt were being opened up, and for many years afterwards, it was possible to plough with horses and mules; nor has fertilizer been necessary for unirrigated prairie wheat at any time. The first trials with superphosphate for Australian wheat were made only in 1885; and fertilizer can hardly be said to have become important in Australia until about twenty years later.

It is significant that the recent adoption of irrigation in Alberta has led to a demand for superphosphate and other fertilizers on the irrigated area. That was, indeed, inevitable.

The present picture of world food exports and imports is essentially an affair of this century: not, as history books give the impression, of the middle of last century. There have been impressive changes in the tableau since 1920 and more recently.

It would be no exaggeration to put the development of the *modern* food problem – the problem, that is, which affects you and me now – alongside the development of the aeroplane. The correspondence is almost point by point. True, there was no power-driven aeroplane of consequence before 1903; and true, that people are now eating the same sorts of foods as people were eating in 1850 or earlier. Nevertheless, the comparison is fairer than may at first appear.

Most of the developments and changes in food-production which are now material having occurred since about 1920, what happened before then is as historic as the biplane, or the first flights across the Atlantic. As will be shown, the industrial unveiling of the crux of the fuel problem in its relation to food – namely, the question whether sulphur or other fuel shall be used in profusion to build food and to maintain its consumption soaring – is within one or two years contemporary with actual flight in heavier-than-air machines.

All the new land is known, nor is there much of it left to be brought under cultivation with or without use of fertilizers. What was new land in 1880, or in 1910 or 1930 is no longer virgin, nor perhaps as productive as it once was. Suns of a few summers beating on ungrassed land soon find less and less organic matter to burn away from the soil's diminishing stock. When there was evermore new land to crop, few minds were troubled. There was always Horace Greeley's advice 'Go West, young man; go West'; and it is worth remembering that this was not simply an injunction to the adventurous at large that they should be bold, but was essentially a hint that if the 'old' land was 'exhausted' after a few years' cropping, there was plenty more to be found after the 'old' had been abandoned to its fate.

Such decrements of fertility occurred on a large scale, and until recent years. They did not matter for a time (or so it seemed). Total yields could be kept up, and export totals increased, by breaking more land. Now, the honeymoons are over; the extravagant newly-weds must once more look to realities, and settle down to economical housekeeping and to husbanding their not illimitable resources, in order to have some hope of providing against a future which – among other things – gives them power to add to their number.

Among assets and resources there is sulphur. This is used in the making of all sorts of things, including paper, some kinds of rayon, pneumatic tyres, and the most important fertilizers. The rise of the automobile industry – and hence a big demand for the two fuels sulphur and petrol – *happened* to begin at about the time the big U.S. deposits of sulphur were first worked on the large scale; but, of course, if sulphur from the deposits known at about the beginning of the century had not come from these U.S. deposits, it could have been made by other means: as, for example, by chemical processes involving consumption of coal. Some sulphur is indeed prepared chemically. The point of these remarks is to be found partly in the fact that sulphur is a fuel, and is obtained only by expenditure of fuel. The fuel used for producing sulphur may be oil, coal, or sulphur itself.

British readers will have noticed the lack of official grasp of the idea that sulphur is a fuel. The various bodies concerned nominally to give advice about use of fuel and power do not include sulphur in their terms of reference. This is true even of the latest such organization – set up in 1953, not long after a sulphur crisis was news. The trouble, I think, is not so much departmentalization of interests as compartmenting of ideas.

The automobile uses three kinds of fuel. Besides the petrol or oil consumed rather wastefully in the engine and transmission, the automobile accounts for a good deal of the world's annual output of sulphur (in its tyres and gaskets). There is the further consumption of fuel for making the carbon black which is another component of tyres. An American estimate[1] rates the automobile very low in the scale of fuel efficiency with respect to what is burnt to provide motive power. The authors of this estimate also put out the statement (quite astonishing at first sight) that to make the few pounds of carbon black needed for the tyres of an ordinary motor-car the amount of fuel customarily consumed has an energy-value equal to about a tenth of the petrol consumed, on an average, in a year's running of the car.

Whether we like it or not, the technically-advanced nations must face the position they are in with respect to fuel. It can be stated briefly: industry, transport, and food-production are all in competition for a limited supply of fossil fuels.

That we have had only one crisis of sulphur-supply this century is due to the fortunate discoveries made at its

1. *Energy Sources: the Wealth of the World.* By Eugene Ayres and C. A. Scarlott. New York and London: McGraw-Hill, 1952.

beginning. The supplies thus revealed were far greater than any previously known; and they yielded enough sulphur to last, without incident, for fifty years.

Until the end of last century, sulphur came from three principal sources: elemental sulphur mined in Sicily and other parts of Italy, or recovered from various chemical processes; and burnable sulphur in sulphide ores. (Much of the latter was wasted in extracting metal from the ores.) With some fluctuations, these sources met all needs up to 1900 or so.

Much the greatest of the uses of sulphur was, and is, in the making of sulphuric acid. The greatest use of sulphur as such is as a vulcanizing agent for rubber. To make sulphuric acid, sulphur has to be burnt; the resulting sulphuric acid is used for purely industrial purposes, and to meet war-time demands, and for manufacture of the most important fertilizer – superphosphate.

As the biggest consumer of sulphur last century, the United Kingdom was directly interested in the then almost exclusively European sources of sulphur. In the 'nineties British capital was invested in the Sicilian mines; and before that, a British warship was once sent in peace-time to point out the importance of not hampering the free flow of sulphur to our paramount industries.

If we leave out of account metals and carboniferous fuels, superphosphate is now the most important single substance produced by chemical manufacture. In 1861 the world's production of superphosphate was about 160,000 tons, nearly all of which was made in Europe. In 1951, more than 27,000,000 tons of superphosphate were made: about half of that having been produced in the United States for United States consumption.

Why is superphosphate so important? Because to make superphosphate is still the readiest way to render the phosphate of phosphate rock (fossil phosphate) available to plants. It is irrelevant that conversion of phosphate rock into superphosphate is the oldest process for making the fossil phosphate plant-available; it is largely irrelevant hat the phosphate of superphosphate is soluble in water. (Superphosphate is often treated with ammonia or in other ways to alter its properties, and so to bring about a decrease in the water-solubility of the contained phosphate, without rendering it appreciably less available to plants.)

Superphosphate was made about 1840 as 'dissolved bones' and other names, by mixing sulphuric acid with degreased bones. When superphosphate was patented by John Bennet Lawes in 1842, neither the material nor its name was new: except that superphosphate was to be made from fossil forms of phosphate. The merit of Lawes's invention lay in his seeing that there would not be enough bones for an expanding agriculture. Lawes made it possible to tap not only the small known fossil resources of phosphate, but – as it turned out – to utilize the immensely greater deposits of rock phosphate which were to be discovered in and after the second half of last century (though not much used until after 1900).

The manufacture of superphosphate is easy. Though mechanization has been adopted, the actual making of superphosphate is in principle the same to-day as it was when superphosphate was made on farms long ago from bones. It consists of simply mixing the ground phosphate with rather less than an equal weight of sulphuric acid containing some water. The mixture heats, and is left for some hours to adjust or 'cure'. The superphosphate is then ready, except for grinding and bagging.

The sulphuric acid will have been made either by burning elemental sulphur (that process is simplest, and demands the smallest outlay of capital) or by burning sulphur of pyrites or other sulphide ores such as zinc blende. This combustion or oxidation of sulphur releases a good deal of energy as heat through reduction of atmospheric oxygen (just as when coal burns). Some of that energy is ultimately applied to the fossil phosphate rock when the rock is mixed with sulphuric acid to unbind the actual phosphate from the substances with which it is combined in the rock. When laid down, the rock phosphate was (almost invariably) of organic origin, and therefore its phosphate could have been used by plants but it was deposited – mostly on the sea floor – out of the way of plants; and in the course of ages it altered to a complex phosphatic material, and the contained phosphate ceased to be readily plant-available, even after fine grinding of the rock.

Many processes are known for rendering the fossil phosphate plant-available once more. They are all expensive in fuel and energy: all of them require complicated processes besides grinding the rock phosphate; moreover, depreciation of the machinery is usually severe, owing to the general requirements of high temperatures and strong acids or both. The making of superphosphate in the manner just outlined is the mildest as well as the simplest of these processes, and therefore holds the field in face of competition. Besides that there is the fact that superphosphate is usually acknowledged as the best of the phosphatic fertilizers purchasable in large quantity. In field experiments, for instance, ordinary superphosphate is normally the standard with which the performance of equal amounts of phosphate in other fertilizers, new or old, is compared.

During the last three decades great industrial demands have been made upon elemental sulphur (especially for rubber tyres) and upon all kinds of burnable sulphur for making sulphuric acid for old and new industrial processes (including production of uranium) and for manufacture of the greatly increased tonnage of superphosphate needed to sustain this century's new mantling of leguminous plants and other crops: not forgetting the newer and more productive varieties of crops such as sugar-beet and cereals.

There have been changes in the sources of sulphur during the last fifty years, but the only one which need be dwelt upon is the effective exploitation since about 1904 of the deposits of elemental sulphur in Texas and Louisiana. When examined before the first World War, these deposits seemed almost illimitable – at the rates of consumption then current; now, it must be said that chemists tend to think of them mostly in the past tense. *Où sont les neiges d'antan?*

These sulphur deposits of the Texas Gulf lie in water-logged ground (in part under the sea) where no man can work and live. Attempts to get the sulphur out all failed, until an ingenious process was developed at the turn of the century by Herman Frasch for melting the sulphur of the richer portions *in situ* and forcing it by pressure to the surface. This yields one of the purest raw materials in commerce.

Admiration for the end-product of the Frasch process and for the ingenuity of the process itself has led to some misapprehensions. Chemical text-books not seldom contrast the American method of winning sulphur with the standard Sicilian processes, which are incapable of being mechanized and for which a rather poor grade of ore is mined by hand. Worse still: the Sicilians roast out their sulphur in old-fashioned stacks called *calcaroni* (or by a regenerative process on the same principle), and actually burn some of the sulphur to provide the necessary heat! This, it is implied, is reprehensible; even, primitive!

Such criticism would not be made except by people whose outlook is that it is quite natural to burn coal (or petroleum), but horrifyingly unnatural, if not uncivilized, to burn sulphur. Yet, what are the poor Sicilians to do to be saved – having no wood or coal at hand, or anything else to use as fuel?

The consumption (waste, some would say) of sulphur in Sicilian *calcaroni* and kilns is about 30 per cent of the sulphur mined. The fuel economics of the Frasch process is obscure. I have seen no helpful figures, except an account of one recently-opened U.S. mine planned to produce 500,000 tons of sulphur a year – more than the entire output of Sicily.

This account gave full details of production of electricity, hot water, and steam needed for a year's operation – but, characteristically, it gave no hint of the fuel which would power all that. Presumably, the fuel came from the near-by oilfields, and was taken for granted. From the data given, and making reasonable assumptions not unfavourable to the process, my friend Dr Frank Rumford calculated that to

extract a ton of sulphur about one-eighth of a ton of oi
would be required. From a hasty point of view, this appear
to be more economical than the Sicilian use of fuel; but it i
not clear how much sulphur is left in the ground – not burnt
but irrecoverable for ever. It seems that the 'stream-lined
modern American process may not be less wasteful of sulphur than the derided Sicilian ones, when all things are
considered.

Even more to the point is this thought. To look upon the
Italians as benighted because they sometimes use sulphur a
fuel is to forget that anyone – in Britain or elsewhere – who
burns sulphur from any source with the object of making
sulphuric acid to convert bones or phosphate rock into
superphosphate is doing precisely the same thing: namely
burning sulphur for the reason that to do so is the easiest
handiest, and cheapest way of obtaining energy for the purpose desired. Superphosphate could be made by burning
coal instead of sulphur (by using coal to roast gypsum: some
sulphuric acid is made in that way); or a fertilizer could be
made, and sometimes is made, by heating phosphate rock
chemically or electrically with coal instead of using any
sulphur-containing ore; but all such substitute-processes
without exception are more difficult and expensive to
operate than is the manufacture of superphosphate by burning elemental sulphur, or sulphur in the appreciably more
expensive source, pyrites.

Under this head, it may be repeated that hydro-electric
power is not a substitute for a reducing agent such as sulphur
or coal in a chemical process. Accounts of the manufacture
of phosphate fertilizers by hydro-electric power in, say, the
Tennessee Valley; and the still more widely-publicized information about schemes to make aluminium (to which
metal the picturesque name of 'frozen electricity' has been
given) – usually omit to provide the dull but necessary
details about the large amounts of carbon consumed.

It takes a chemical to make a chemical; and, until it
becomes possible to subvert the recognized laws of chemistry, there will always be the necessity to use something more

han the calculated minimum of coal in making phosphorus (for phosphoric acid and phosphate fertilizers) or aluminium or any other structurally useful metal, or compound (except nitric acid, page 138), even if the power for the furnace is all derived from water.

A not inconsiderable part of the world's phosphatic fertilizer is basic slag. This is a by-product of steel-making, so that it might be fair not to debit any fuel against production of the slag. Much of the basic slag yielded by modern processes of steel manufacture is of inferior quality as fertilizer, since the phosphate in it is not of high availability to plants. A large proportion of the basic slag sold as fertilizer in the United Kingdom is imported.

Without going further into details, it can be accepted (I suggest) that the future expansion of phosphatic fertilizer output, and consequently of cultivation of legumes and of production of protein foods and their desirable accompaniments, must lie with superphosphate made by ignition of sulphur, or else with an equivalent tonnage of fertilizer constrained to have a similarly high proportion of plant-available phosphate and made by consuming at least the same weight of coal as the sulphur it is intended to replace.

It can be seen how lucky the automobile industries and the agricultures of to-day have been in finding their needs for sulphur met in such gratifying fashion by the new sources of sulphur exploited as they grew rapidly to supply transport and meat and milk and edible oil and cheese and butter. How long this fortunate conjuncture of circumstances will last is hardly for me alone to guess. It was of the nature of a fluke, and our food supplies have benefited from it.

The reader may like to compare the tonnages given and implied in this chapter with the figures on pages 148 and 151, which relate to about four million tons of superphosphate.

Incidentally, the half-million tons of pure sulphur to come each year from the new U.S. mine mentioned above will fall short of the annual increase in U.S. consumption of sulphur for fertilizers that is expected to supervene within the three years after 1953. The acreage under cotton in the U.S. is not expected to be greatly enlarged, but higher yields are hoped for from more generous treatment and better management of all U.S. crops. The *increase* in consumption of sulphur for fertilizers, by that number of people which brings the U.S. population to about 180,000,000, will be about twice the annual output of sulphur for all belligerent and peaceful uses by the whole world of sixty or seventy years ago.

That being so, nations less amply equipped industrially than the United States is, and having less spare land and fewer mineral resources than that seemingly well-endowed country, might ask themselves how they are going to allocate their disposable fuels between the demands of industry, transport, and agriculture so as to feed the populations of the immediate future.

Because sulphuric acid enters into the making of so many things, it has long been supposed that the output or consumption (practically the same thing, since sulphuric acid does not travel well) of sulphuric acid is a measure of a country's 'civilization' – meaning its industrial development. The idea grew up last century – that is, before the Second Agricultural Revolution had made it appear that one of the most important uses of sulphuric acid has small direct connexion with trade in manufactured goods of the kinds usually thought of as industrial or technical products. It is interesting to review the position in the light of to-day's actualities.

Dr Alexander Fleck has shown[1] that in the United Kingdom the growth of consumption of sulphuric acid has for about a century corresponded closely to total volume of manufactures. In the United States, however, there has been a very recent tendency for the annual output of sulphuric acid to grow faster than the volume of general industrial output. There can be little doubt that this modern divergence is an expression of the late 'discovery' of the usefulness of fertilizers – notably superphosphate – by the United States as a whole.

1. A. Fleck: *Chemistry and Industry*, No. 49, 1952, pp. 1184–93.

The recent upsurge in cultivation of alfalfa, soybeans, and other leguminous crops in the United States has led to a great new demand for superphosphate.

This demand can be expected to grow still further as a consequence of the cultivation of hybrid maize and other higher-yielding crops, coupled with more careful attention to productivity of soil than was given before (say) 1930, or is being effectively given now. Immense as the expansion of general industrial output of the United States has been in the last few years (with a probably parallel demand for sulphuric acid for strictly industrial purposes) the consumption of sulphur for fertilizers has been boosted still more enormously by the new cropping.

If we put 1 for the steady ratio between output of sulphuric acid and general volume of manufactures which prevails in the United Kingdom and held until about 1935 in the United States, the present ratio in the United States might be almost 1⅓; but in Australia and New Zealand the ratio would be (I estimate) about 9 and 19 respectively.

New Zealand has the highest per capita consumption of sulphuric acid in the world; Australia comes second, and the United States third. The respective figures, in pounds per annum, are about 230, 180, 120. New Zealand has hardly any manufactures except of superphosphate, and Australia has not many either. The figures refer, conventionally, to 100 per cent sulphuric acid (though that is not the form in which most of the acid is produced) and can be expressed as sulphur by dividing by three.

The relatively high U.S. consumption of sulphur for sulphuric acid is largely accounted for by the intensity of her manufactures, but the higher figures for New Zealand and Australia represent, almost entirely, superphosphate and a little sulphate of ammonia; the latter being used on arable crops. New Zealand has the most productive grassland in the world. Most of the New Zealand superphosphate goes on to grassland, and practically all of it represents animal products: *i.e.*, protein foods, wool, and butter for the peoples of the United Kingdom, in essence.

New Zealand seems to have no sulphur ores of any kind, and all her annual consumption of 80,000 tons of sulphur has been imported, as sulphur; mostly from the United States. Australia also relied heavily on U.S. sulphur, but is beginning to use her native pyrites. If New Zealand failed to get the equivalent of about 100 tons of sulphur per day, or if the population of New Zealand were to double, there might be no New Zealand animal products on United Kingdom breakfast and dinner tables; but, of course, we should continue to receive wool if we wanted it.

CHAPTER SIXTEEN

Notes on Irrigation

> The antient chemists thought that salts were formed of water and earth; but they admitted a third principle, . . . – *A Manual of Chemistry*

THROUGHOUT this book I have not relied on theory but have tried to show that continued practice agrees satisfactorily with what is known to scientists about the nutritional needs of plants and animals and about the behaviour of the soil and its inhabitants. It will be self-evident that practice in farming cannot be continued long on a given site unless the sum and detail of its operations are in harmony with the facts of life.

Prolonged further exposition along those lines might interest only agriculturists. I therefore take only two matters for expansion – having chosen those two from among the favourite prescriptions for getting higher yields or obtaining a larger total of food. I give brief remarks here about irrigation. The next Chapter offers a longer discussion of some aspects of higher yields from other devices. It would be possible to discuss almost any prescription for greater output of food, and to reach a devastating conclusion.

Irrigation is almost invariably thought to be synonymous with giving water to overcome a deficiency of moisture during a dry season or in a dry place. In fact there are several other purposes which can be discharged by irrigation. The common feature of all forms of irrigation is giving water – not lack of moisture. The water may be a carrier-away of salts (as during reclamation of saline dry land or of land being reclaimed from the sea); it may be a carrier of some useful substance, such as silt or dissolved fertilizer or oxygen; or it may be a carrier of warmth. This last function is exemplified in the English water-meadows and catch-meadows and still better in the North Italian *marcite* or *prati iemale* ('winter-meadows'). These irrigated meadows are all watered in winter, when there is plenty of rain. Sometimes two functions of irrigation are combined, as by irrigation of crops with dilute fertilizer-solution during a dry period in spring; or the same installation can have different functions at different times of the year.

The golden rule for all forms of irrigation is without exception the same: it is that efficient drainage must precede and accompany the letting-in of water. The good drainage may be natural; if good permeability and drainage do not exist, they must be provided and maintained. Whenever that rule is not observed, various kinds of trouble quickly follow singly or together, and crops fail, or grow with difficulty. Troubles often arise from presence of excessive water in the soil (not enough air); or an accumulation of salts may supervene from various causes, and this has the effect – however curious it may appear – of reducing the amount of water *available* to plants. After all, it is the amount of water that is available to the crop which matters – not the volume which is led on!

I can give here no more than a hint about this important point of reduction of availability of irrigation-water through soil salinity, whether the salinity is natural or is artificially induced by careless irrigation or other means.

Suppose you thought of using sea-water for irrigating land. You would at once realize that it is too salt to permit growth of most land plants. Therefore you would plan to reduce the salinity either by removing the salts, or by adding as much fresh water as possible. It would be necessary to mix the sea-water with at least ten or twenty times its own volume of fresh water. Sea-water contains 3½ per cent of salts, and 96½ per cent of water: yet all that water is unavailable to crop plants. The Dead Sea contains no more than 30 per cent of salts and about two-thirds of it is water, none of which is available for any vital purpose by any high form of life

Notes on Irrigation

Irrigation schemes promoted by white men for hot countries have failed within our own century from such causes as over-irrigation and accumulation of salts (the two often going together); and a very considerable area of land meant to benefit from irrigation has been spoiled by it for crop-production. The traditional system of the lower Nile and some well-thought-out modern schemes continue in being. The lesson of the Nile is that it supplies perennial water of good quality, and the lower Nile, at least, offers a perfect drainage system. Here, as elsewhere in durable agricultures, legumes are prominent.

Egyptian agriculture and some other agricultures of hot countries in the northern hemisphere are centred on *bersim* or berseem (Alexandrian clover: *Trifolium alexandrinum*).

Like any other form of farming, cultivation under irrigation involves a price to pay for the disturbance of natural balance. Irrigation is no automatic means towards extra food; nor are the capital and maintenance charges of the engineering works its only cost. As with other disequilibrations set up by farming, the penultimate brunt must be suffered either by fuel or by man himself. There seems to be no other alternative. If any new level of plant output reached, for example, through irrigation, is not sustained by fuel, the consequences must be borne by the human population.

That conclusion is not widely accessible to a public which hears or reads stories of vast engineering works about which pictures and propaganda are put out by the Marvellous Job boys employed by some collection of initials. Their emphasis is on great effort while they make play with millions of acres, so that one might gather that Nature has again been mastered and to a significant extent.

A remedy for that sort of thing is to count one's snows, including the head of the Nile. Political control of snow seems a silly thing to fight about; yet a study of a relief map of the world will show that there is small scope for expansion of irrigation, and will suggest the realist explanation for at least one international dispute.

The maximum which can be irrigated in Canada is about a million acres in Alberta. Schemes in Australia which include extensive diversion of rivers will irrigate perhaps three million acres (4,700 square miles); and that, for Australia, is the practical end of irrigation, since there is no more snow. Altogether, the irrigable land of the two countries could feed at most two and a half million people.

A demand for fertilizers already set up by irrigation in Alberta can be expected to grow. If the resulting fuel-demand is pooh-poohed as too small to be of much consequence, let it be remembered that the numbers of people are in the same proportion.

CHAPTER SEVENTEEN

An Example Worked Out

Alice had got so much into the way of expecting nothing but out-of-the-way things to happen, that it seemed quite dull and stupid for life to go on in the common way. – *Alice's Adventures in Wonderland*

THOSE countries which have most vigorously taken up and evolved better varieties of crops and animals or are in course of adopting them are precisely those which have the best systems of education, and have or are proposing to have programmes of intensive use of fertilizers.

Thus Denmark, famous for the high and uniform quality of its pigs and bacon backed by first-class dairying and a system of Folk Schools disseminating general culture; Holland, whose agricultural and horticultural skill is almost proverbial; and Belgium, which shares with Holland the distinctions of having very high yields of crops and possessing a surprisingly large number of Universities and Technical High Schools (of University rank) for so small a population: these countries and Sweden are playing active parts in disseminating new varieties of crops and in establishing efficient races of farm livestock. With Germany, they are the heaviest *per capita* users of fertilizers in Europe.

Nor are New Zealand and Australia lacking in education. New Zealand benefited from her kindly climate to establish the highest-yielding pastures in the world, but she had to start from scratch not much more than fifty years ago (her whole 'white' history is barely 110 years long); as did Australia in emerging from the ranching or pastoral stage.

Where else is the level of rural education high, or about to be raised? Need I mention Ayrshire cows, or the agriculture of the Lothians? The big test will come in the United

States, which in the last sixty years has enthusiastically taken up several new kinds of crops. The United States has a chequered history of management of its soils, forests, crops and animals; until recently, expansion of total yield had mostly been obtained from soil exploitation – by adding to the soil area rather than adding something to the soil. Education of its farmers and better management of the land can obviously do a lot more yet for that country of large (though still limited) scope.

It is important to understand that fertilizers and adoption of new or better crops and animals are but *parts* of management of the soil and the farm. To use any one of these parts by itself will almost certainly fail to achieve a significant and useful result; to use two will, exceptionally, succeed (and is therefore in the nature of a gamble). Good and lasting results can come only from a balanced combination of all three: more tricky than it appears, since fertilizers must be balanced with each other and with lime, and foods both for animals and man must be balanced according to season and climate and in content of nutrients and minerals.

To sow seeds of any kind of plant except in a habitat known to be suitable is to invite the grossest kind of failure. If the habitat is not already suitable and in condition to remain so, it must be adjusted by every appropriate means. Fertilizers and the rest are no more than means to an end. It would be almost as silly to sow fertilizers and to hope for a crop (*something* is sure to grow!) as to sow seeds of high-yielding plants without making provision for the additional drafts which the hoped-for higher yields will make upon the nutrients (and sometimes, water) of the soil all over the farm.

One example will help to show the truth of this. The example centres upon our own British crop of wild white clover. This is not a foreigner. Last century and earlier it occurred plentifully in the truly wild condition in Kent and other parts of southern England, and still does so. It was not originally a new variety in the sense that hybrid maize and the newer varieties of artificially-bred grasses are new. Its conscious encouragement this century throughout the United Kingdom and every other place where grass, as we understand it, grows does pointedly indicate the need for combining management, fertilizers, and seed into a productive system. It also suggests the responsibilities which ensue from adoption of any new and highly-productive system of management of land.

The scene is Cockle Park in Northumberland. This is an experimental station belonging to King's College, University of Durham, and is some 400 feet above sea level. Not quite 'marginal' land, some of its fields can be cropped, though most are left down to grass of poor quality in 1896, when the story starts. Dr William Somerville, the first Professor of Agriculture in the University of Durham, is the Scientific Director.

Professor Somerville begins trials in treating the pastures with basic slag (a phosphatic fertilizer which contains available lime). He uses the slag at the rate of 10 cwt or so per acre for a first dressing on land which has had hardly a taste of fertilizer before; that initial dressing is and will be followed up by repeat dressings of 5 cwt of slag per acre every third year. He finds that the phosphate and lime in the slag appreciably encourage growth of the wild white clover which has hitherto just managed to exist sporadically throughout the permanent (unploughed) grassland.

Nothing much is yet known about the bacteriology of legumes or about soil microbiology as a whole. The science is still being born, and Dr Somerville makes no special use of it, though he is aware of the role of biological nitrogen-fixation in relation to legumes. As a practical man, he notices that the better growth of wild white clover is accompanied by better growth of grasses; not only does the herbage grow more rapidly, yielding more bulk in a given time, but the more nutritious grasses are more in evidence: the pastures are improving.

More sheep per acre are required to stock the pastures properly. This is a definite result: greater carrying capacity for animals through use of basic slag. Now Somerville begins to try potassium salts, and liming (to supply more lime than the basic slag alone does), and nitrogenous fertilizers: all in various combinations with basic slag, and superphosphate too.

In 1899 he is succeeded by Professor Thomas Middleton, who continues along the same lines. He brings cattle to graze with the sheep. This mixed and heavier grazing of cattle and sheep is in accordance with the growth of the greatly improved, better-yielding, and more closely fenced pastures, manured and treated in what has been found to be the most productive scheme of manuring – namely, a balanced one. Some of the pastures are ploughed up, manured appropriately, and sown with cereals and other crops to give keep (food) in winter for the extra livestock carried in summer on the rest of the grassland, now rich in wild white clover and good, nutritious grasses. More and better hay is taken, too.

In 1902 there is another change: Douglas Alston Gilchrist comes as Professor of Agriculture and Director of the farm. Gilchrist is a Scot, born in 1859 to a farmer at Bothwell, near Glasgow. At about the age of 26 he had attended classes in agriculture and science at the West of Scotland Technical College (the parent body of my College and of the Royal Technical College) and went on to take a degree in agriculture at Edinburgh – then the only British University with a degree course in agriculture. When little more than 30 he was selected to open agricultural instruction at Bangor in North Wales; a few years afterwards he did similar service for Reading; and in 1902 he takes the Chair of Agriculture at Newcastle-on-Tyne.

Gilchrist inherits the system of management evolved by Somerville and Middleton. It is complete, as far as it can be made for permanent grass such as that which abounds all over England and Wales and is not unknown in Scotland and is already populated, however thinly, with 'wild white'.

But supposing the old grassland – in Northumberland or Leicestershire or elsewhere – is once ploughed up? Will the valuable clover come again for many years – won't the land be less productive if, say, the sacred permanent grass of the English midlands is ever ploughed? Or what about Scotland, where they have a habit of ploughing up most of the grass in rotation every few years, thus stocking the soil with organic matter for the arable crops which follow, but having rather poor grass in-between-times?

(In 1896 or 1902, please remember, the days of good general treatment of grass, and the attitude of looking upon grass as a cultivable and improvable crop, had still to come. The English graziers who had good clovery grass – and knew it – were afraid to plough it in case they never saw the like of it again on their land. Gilchrist was aware that there was hardly any grass in Scotland as good as that upon which he entered at Cockle Park.)

There was a saying in England:

> To break a pasture makes a man;
> To make a pasture breaks a man.

This meant that to cash in on the stored organic matter and nitrogen of the soil of the better permanent pastures was to have good crops of 'roots' (turnips, mangolds, potatoes) and corn for a few years; but it would be a slow and unremunerative business to restore the pasture to anything like its former productive condition by sowing the grasses and short-lived clovers of kinds obtainable before Gilchrist found what to do.

There was in Stirlingshire at about the turn of the century a brilliant farmer and public-spirited man who interested himself in making grassland productive. He was Sir David Wilson, a land-owner and chemist who made his own analyses of the produce of his experiments; and he showed before 1885 that he had a grasp of all the essential features of productivity of grassland. He lacked only one thing: a perennial pasture-legume. The red and white clovers purchasable before 1906 were no more than hay plants, fit to grow a year or two in the exclusively arable rotations typical of the First Agricultural Revolution, but they were not able to confer benefit for several years under grazing conditions.

Good as the clovery grazing lands of Kent and Leicestershire might be, and as the pastures of Cockle Park had become, they had to stay where they were; they could not be spread about. Gilchrist realized that while pastures were not portable, the seed of (perennial) wild white clover was. In 1886 near Chester, on land forming part of the Hawarden estate of W. E. Gladstone (the famous Liberal leader), Gilchrist had seen trials of sowing wild white clover. The seed had been obtained from the Weald of Kent, where it was not entirely unknown locally. A few years later he tried sowing the seed himself not far from Bangor, and again while he was at Reading. But in none of these instances had the clover taken well; nor had the surviving plants flourished. Perhaps these early failures had been 'explained' by saying that Kentish seed wouldn't do well anywhere except in Kent? The failures might even have been taken as proving that slender hypothesis.

At Cockle Park, Gilchrist had the clue. He wrote later: 'It is probable that in these early trials the clover would have done much better if basic slag had been given as well.' As with the development of miscellaneous microbes in associations and communities, it was *conditions* of environment and living – not the presence of species – which decided whether an introduced species or one better suited to the prevailing conditions was to flourish or become dominant in competition with the numerous other species present and possible. The conditions at Chester and elsewhere had not suited clover: the soil was too acid, at least; so the local grasses and weeds which appreciated acid conditions went on growing, but the legume, more particular and requiring lime and ample phosphate, could do no more than make a beginning at growing.

Kentish seed failed not because North Wales or Chester was not Kent, but because the sites of the early trials of sowing wild white clover had not been prepared to be as suited for that clover as the well-treated Kentish soil was. So long as soil in Wales or England or anywhere else was acid and short of lime and phosphate, no clover (whether labelled 'perennial' or not) had a chance to flourish and persist.

To make sure of having a perennial clover – or any other plant – in grassland it is not enough to sow the right kind of seed, or to sow basic slag (with or without potassium salts and the rest). To do no more than sow the seed is to invite failure, even should the climate be nominally favourable; sowing fertilizers containing phosphate and lime would greatly encourage clover on land which already bore it, but would not help much after the land had borne arable crops for some years.

In 1906 Gilchrist induced seed merchants to begin to offer seed of wild white clover; and he began to preach the doctrine of including the seed of that clover in every mixture of seeds intended for sowing-down land to grass after it has been arable. Pastures were not portable; but 'wild white' and other kinds of seed were.

In order to re-establish a good pasture after it had been ploughed, there was no longer the need to depend upon chance, or upon sowing the kinds of seeds previously judged, without real knowledge, to be suitable. It was, however, essential that the ploughed land should first be manured well with basic slag and other fertilizers to give both the clover and the grasses the mineral nutrients they required: they then had an opportunity to grow strongly from the beginning of their lives. Afterwards, it would be necessary to manage the new (or old) grassland properly with a due regard to what livestock it could carry. Under those conditions the good and nutritious plants would persist and cover the ground, so that inferior plants which might come in adventitiously would find neither room nor conditions for their growth.

It was the system of management evolved by Somerville and Middleton that Gilchrist recommended, together with the sowing of wild white clover seed in chief. Along with that seed went a considered mixture of grasses selected by Gilchrist. This grass-clover mixture became famous as the Cockle Park seeds mixture for long leys or permanent grass.

In conjunction with the advice about manuring to establish conditions, and grazing management to maintain the right conditions, this mixture of seeds centred upon a legume offered a prescription for better husbandry all over the farm. The Second Agricultural Revolution had broken out in Britain.

This made it possible to plough up old grassland not only with impunity but with positive benefit. The new grassland – better manured and managed, and with clover well established – could almost always be made more productive than the old permanent grass. Scotland had already discovered that, in part; but henceforth the frequently-ploughed Scottish grassland could be – and now often is – as well endowed with protein-rich perennial clover as any of the old English or Welsh permanent pastures plentifully besprinkled with unsown 'wild white'.

The advantages did not confine themselves to the pastures. The benefits extended all over the farm. For one thing, to the extent that the land was better provided with lime and phosphate carried to the grazing animals or in hay by the clovers and better grasses, the animals were more numerous and thriftier: they avoided deficiencies of calcium and phosphate and could give more milk, for example. Not that the millennium had arrived – I do not wish to paint too rosy a picture or to suggest that all was learnt at once. There is still room for improvement in manuring and management of grassland, at least; but every farmer now has wild white clover at his disposal – whatever the degree of use he makes of it.

We reap to-day the benefit of all such work. For instance: during World War II, old grassland could be ploughed up to bear crops for direct human consumption, wherever the climate enabled such crops to be grown. The great increase in output of fresh milk in the United Kingdom in recent years is in large part a consequence of the work of Somerville, Middleton, and Gilchrist. Yes, of course there have been more cows, and better-managed cows, and more farmers going in for dairying, to provide that extra milk;

but the cows had to be fed with good grass or some other source of protein (mostly imported if it wasn't grass). And were not the cows and bullocks of dairy and beef breeds bound to have extra phosphate and lime not simply to keep the clovers and other crops growing, but to make good the heavy drain of phosphate and lime taken from the soil by the animals, and sent away to the cities in milk and meat and bones to nourish the babies and bigger citizens?

The Cockle Park workers were not alone in showing the responsiveness of clovers and other legumes to appropriate manurial treatment. Their great merit was that they did not stay on any one point, but worked out a complete system of management. These are some figures of livestock carried at Cockle Park on a constant area of some 420 acres of land not outstandingly blessed by climate and situation:

Year	Horses	Cattle	Sheep
1896	5	12	124
1939	7	152	279
1949	7	150	511

Gilchrist expanded the benefits beyond northern England. His highly practical advocacy of wild white clover included in a suitable seeds mixture, and of the need for lime and phosphate to go before it, has probably done more to make 'two blades of grass grow where but one grew before' than any other single innovation in Britain this century. To say this is not to forget that he was following up ten years' work and the demonstrations of manuring and management by his predecessors.

The effects obtained by Somerville and Middleton had been clearly demonstrated within a few years on a farm within their own control. That was fortunate for them, since it enabled their results to be credited to them. The merits of Gilchrist's introduction of wild white clover seed into commerce could not be appreciated for many years, until farmers in general had made it their practice to include some 'wild white' in seed mixtures they ordered. Nor could it be said at any one time that 'wild white' had arrived. The seed was taken up slowly over decades. Only after 1935 (I hazard a guess) may it be said to have been generally used in the United Kingdom; and war-time circumstances (of the second World War, that is; the first World War did no more than give impetus to its use) finally established it as indispensable. Nowadays 'wild white' is an ingredient of every mixture of seeds intended for sowing-down agricultural grass meant to last more than a year or two.

By 1935 Gilchrist and the part he played had been forgotten. There had been knighthoods for Somerville and Middleton; but Gilchrist died at Newcastle in 1927 with his achievements unrecognized except by a few specialists. Those achievements were not of the kind which produces a gadget or attracts political notice.

Wild white clover is not spectacular. Being grown always with grass, it is never registered as a separate crop, and thus its expansion attracts no recording in official statistics. Gilchrist had produced no sudden novelty, and while he lived there was nothing very tangible to demonstrate as a result of his efforts. He had only assisted British farming to become more productive, and to reach new productive heights more easily and with flexibility. He had exploded the old saying about 'breaking a pasture', and had turned it inside out: making it possible to cover the British Isles with a leguminous plant for the first time, and ensuring that farmers knew how to treat that plant so that it would endow British farming with new potentialities. Gilchrist had the misfortune to be born, or possibly to work, at the wrong time: during the early years of this century there was a strong current of opinion that the best thing to do with British farming was to give it decent burial.

During World War II we benefited appreciably from our heightened potentials of domestic protein. This statement may seem incongruous, since meat was scarce, and milk was rationed. During most of that war we had no assistance in imports from Holland and Denmark, as in the first World War; and there was restricted importation of feeding-stuffs. The enforced degree of reliance upon home-produced meat and milk led to better farming in which domestic legumes were prominent. If the status of domestic legumes had not been improved – and in particular if grassland of all kinds had not already been to some extent limed and manured in accordance with the teachings of agricultural science – the situation would have been much worse. We benefited, in short, not only from active prosecution of the work of those who taught the virtues of lime and phosphate and legumes under war-time stress, but also from the work of those whose advice was taken, albeit slowly, in the years from about 1900 up to 1939. We still profit and shall continue to benefit up to the limit of cultivation of clovers in our pastures and meadows with still better management and provision of lime, phosphate, and potassium. That limit has not yet been reached.

It is time this story was told. Not only to recall the work and names of the pioneers of Rothamsted and Cockle Park and other centres where the search for the principles of soil management has been pursued: but chiefly, here, to tell of the need for whole management of the soil if the land is to produce more. Dull and dreich a description of such operations of soil treatment may be to many; and few writers and publishers dare inflict the details, or the principles, upon the public. Misconceptions and partial truths prevail because unspectacular fundamentals are often eliminated from public discussions about food-production.

In February 1953 under the heading 'Science on the Screen' the *Radio Times* mentioned a film showing how 'bigger and better grasses' have been developed 'so that a given acreage can support four times as many animals as before.' (This statement made an agriculturist gasp, when I quoted it.) Part of such an increase could be obtained from better manuring and management without the 'bigger' grasses; but, given those improved grasses, the threefold increase cannot be reached, even temporarily, without fertilizers and management to correspond to their needs – and the needs of the additional animals.

No doubt the film mentioned those points; but it is terribly misleading to concentrate publicity on the one item of managing food-production that happens to be photogenic, and to ignore the rest if 'it won't make a picture' or, a paragraph. The large section of the public not already informed about farming is apt to believe that there is no real need to worry about food. Rather will it blame farmers, or a vague 'somebody', for not being progressive enough to adopt modern devices forthwith, so as to double food-pro-tion at a stroke, or at worst within a year or two. If science has brought out new varieties which will give a fourfold increase here or a 40 per cent increase there, then why aren't they grown to give the extra production so badly needed?

The British summer is short (about four months in the Scottish Highlands; a little longer on the Welsh hills). The grazing season in Sussex might be as long as eight months, or a little more. If we keep twice as many animals on grass, how are they to be fed in winter except from additional turnips and corn and hay taken from the lowlands? Does that not mean ploughing up more grass in the sheltered parts, and re-sowing it; and a crop of other problems? In New Zealand, dairy and other cattle can be kept out at grass for twelve months in the year; but hardly anywhere else in the world can that be done. To grow mangolds or turnips is no simple matter while labour is scarce on the farm. In the greater part of the farming United States winter-keep is provided by maize silage, of which the production lends itself to mechanization; but our indifferent climate denies us that resource, as it does many other things. Fig. 7 (page 227) suggests our limitations in Britain. Gilchrist's wild white is our best bet, along with fertilizers and good management for all the crops in a rotation.

The Cockle Park experiments attracted world-wide notice. Their lessons were tried out, and found to be good, in grass-growing regions overseas. Clovers and other nutritious plants were already there, thanks to settlers from Europe; and the grassland plants responded to lime and phosphate and better management in New Zealand and

Ontario and New York State and Ohio and Kentucky and Victoria and Tasmania.

In New Zealand, for instance, grassland was dressed for many years principally with basic slag until that gave way to superphosphate and liming. In New Zealand and Australia the word 'top-dressing' is synonymous with applying superphosphate to grass: neither the crop nor the fertilizer requires to be specified when that word is used.

Then, New Zealand began exporting the seed of her own semi-cultivated 'wild white' to satisfy the expanding demands of the United Kingdom and other European countries for seed of that invaluable plant. Here are a few figures of New Zealand production (mostly from the South Island) of the seed for home use and export:

Season	Acres	Pounds per acre
1937–8	4,800	115
1940–1	20,140	116
1946–7	48,000	124

The 48,000 acres gave enough seed to sow twelve million acres (nearly 20,000 square miles) of grassland, if the 'wild white' was included at the usual rate of about ½ lb. per acre in a seeds mixture.

Mutton is not the only product of Canterbury, N.Z.; but it is no accident that mutton and clovers go together. One kind of high production depends on the other.

The dates are typical in suggesting how recent is the upsurge of effective demand for 'wild white' and other legumes for improvement of grassland and for other purposes in connexion with sustentation of more livestock so as to have much more protein edible by man; and of a demand for more fuel for use in modern intensive agriculture.

Transitional

HUMPTY DUMPTY

... she had read several nice little stories about children who had got burnt, and eaten up by wild beasts and other unpleasant things, all because they *would* not remember the simple rules their friends had taught them: such as, that a red-hot poker will burn you if you hold it too long; and that if you cut your finger *very* deeply with a knife, it usually bleeds; – *Alice's Adventures in Wonderland*

'You couldn't deny that, even if you tried with both hands.'

'I don't deny things with my *hands*,' Alice objected.

'Nobody said you did,' said the Red Queen. 'I said you couldn't if you tried.' – *Through the Looking-Glass*

PASSENGERS in a jet aeroplane may believe themselves t be doing the most modern thing. They may feel that spac has been conquered and that they share somehow in th triumph. Not so stupid or so elated as to believe that gravit is being 'defied', each one of the occupants will know tha the plane is kept up (is in equilibrium) because the engines ar supplying power to keep it so, and that fuel is being burn for the purpose. If the fuel-supply fails, there will be n alternative to coming down to earth or whatever is beneath

The passengers cannot change into a biplane in mid-air It will be immaterial whether the fuel has failed to reach th engines because there was no more fuel on board, or becaus there was fuel which for some reason could not be consumed It will not matter that the machine has not reached th destination which the passengers expected to reach Pleasant or unpleasant, equilibrium there must be betwee the machine and the forces acting upon it; and unless tha equilibrium is artificially constrained by use of fuel i specially-designed machinery, it will be at the lowest leve attainable under the operation of natural forces.

If this analogy seems puerile and trashy, and not worth putting into print, that appearance arises simply because every literate person knows what keeps aeroplanes up, and hardly anyone knows what maintains food-supplies in an age contemporary with aeroplanes. Everybody knows that the one depends upon fuel.

It is significant that nobody, so far as I am aware, has suggested that hydro-electric power would make air travel cheaper or more accessible; nor has anyone put forward the notion that having better-designed shapes or varieties of aeroplanes would enable travel by air to become independent of consumption of fuel. Agricultural chemistry is an economically depressed subject offering few material rewards to its practitioners, and far fewer platforms than are available to, say, aeronautical engineers; with a result that very few people are awake to the fact that fuel is as essential to food-production on a present-day scale as it is to the continued existence of airlines.

Many people will be quite sure that this last remark, also, is an exaggeration. Doubters will be inclined to recall the times when there were no aeroplanes, and people were fed somehow. Yes; without using fuel in food-production, people were fed: but there were many fewer people.

There is also the comfortable riposte: 'Well, things probably won't be as bad as he says. Anyhow, if the worst does happen, we can always take up a few acres somewhere and live on the land. We can grow our own food, at a pinch.'

As regards the first part, I am not prophesying that things *will* be bad, or that a given result will inevitably occur on or about a certain date. Prophecy is not my job. I am a chemist. All I do is write and say: 'Dear Madam or Sir – Unless . . .'; not '—Inevitably . . .' My business now is merely to inform you: by making as clear as possible the conditions underlying that '—Unless . . .'

If you insist on brushing aside the implications of fuel-usage in modern agriculture, I do suggest that the idea of escape into the country for yourself and your family may be delusory. If you are a Frenchman or Frenchwoman, the idea may be sound – so long as your frontiers remain inviolate – because there are possibly hectares enough. But if you are a citizen of some other country, there may not be enough land to cultivate at, or a little above, what Lawes called its 'natural' level. And, if there isn't enough land, do you really suppose that many people will be left in peace to use a spade for their own sole benefit? Might you not have to stand guard over the gates and fences, *par exemple*?

About some such situation, though without explicit reference to fuel, Sir Winston Churchill made the following remarks at the fiftieth-anniversary celebrations of the Massachusetts Institute of Technology: '. . . the consequences would be very unpleasant, because it is quite certain that mankind would not agree to starve equally, and there might be some very sharp disagreement about how the last crust was to be shared.'

A spade in one hand and a club in the other, and neither of them trumps? (Not a *shot-gun* with the spade, remember; unless you and your fellows have given priority, as you might, to fuel for cartridges over fuel for greater production.)

'Ah, now you *have* gone too far!' (do you say?). 'That is funny; but it's plumb ridiculous to suggest that things will ever get that far!' I don't suggest exactly how far things will go. What I urge is that the distance things will and must go, in relation to food-production, will depend on allocations of fuel in relation to food-production. Just that. Comfortably vague in a way, but exact in terms of fundamentals of food to-day.

As a scientist, I can put the principle of food-production before you. It is unnecessary for me to go further into considerations which you can elaborate as well as, or better than, I can – once you have had your attention drawn to them.

In this book there is some repetition. It may have annoyed you. Its purpose is to make some of the later chapters comprehensible singly or in small groups, so that the reader does not have to carry all the unfamiliar preceding matter in his head while recapitulating, for instance. It being impracticable to cater for all degrees of familiarity with my subject, I have aimed at instructing the urban reader, likewise the country gentleman: both of whom may be assumed to have some knowledge of the practices of agriculture, while neither may be supposed to know anything about the principles of soil chemistry or soil microbiology upon which agriculture is based.

The difficulty in exposition thus arises from the inability to assume that anything about which I write is as well known to anyone, except a handful of agricultural chemists, as are the leading facts about aeroplanes. If this were a book setting forth the development of the aeroplane, it would be ridiculously unnecessary to stress – as I have had to stress – the necessity of fuel. Except for very young readers, it would hardly be necessary to mention the fact that unless the structure of a balloon or an aeroplane has a force continuously applied to it, it will lose height and will ultimately reach the ground.

Some of the principal dates in aeronautics would be known to you before you sat down to read; thus the dates of the first heavier-than-air flight in a power machine, of the first crossing of the English Channel or the Alps or the Atlantic, and so on, would occur to you. If you were an American, you might be believing that the first crossing of the Atlantic by plane was made by an American in 1927; but that inaccuracy of period would not matter greatly, since you would be right within ten years. Without having been in an Air Force, you would also be likely to know more or less correctly a large number of other facts about aviation; and terms, such as thrust and wind-tunnel and stream-line and turbine, would convey something to you, perhaps not with entire accuracy but sufficiently well for you to follow an explanation which depended upon amplifying them. You might be under the impression that *stream-line* is a modern American importation into English, but we could soon put that right.[1] Meeting each other half-way, we could make a team.

In enlightening you about food-production, my difficulty – to continue the aeronautical analogy – is that I cannot safely assume that you recognize the difference between a balloon and an aeroplane. (They both need fuel; that is almost the only common point.)

The principles which operate to keep each up are very different; I would have (so to speak) to spend a lot of space in expounding the differences between density and thrust, and their modes of operation, before I could hope to give an intelligible and unclouded presentation of a modern aeroplane. Some of this preliminary matter must seem dull; and I think hardly anyone would attempt to cover the principles of operation of balloon and aeroplane in one book nowadays; still less readily would one engage to do so for readers who had not seen either and were not on familiar terms with the gist of present-day aeronautics.

I have assumed that you know a little about leguminous crops and rotation; but, without insult to your intelligence, I think there is not much else I can take for granted. You will be aware of the relative importance of sea-planes and wheeled planes; but I doubt whether you are aware of the relative importance of legumes and non-legumes for your food. I assume that, like the people to whom Crookes was the prophet of wisdom, you are wheat-conscious or maize-conscious in the sense of believing those cereals to be the most important food crops (as they are the most important in world *trade*). While your information and beliefs about aeroplanes may be sketchy, it is not likely to be seriously wrong in the sense of embracing the highly improbable or having had values seriously misplaced. Can we say as much about your notions of the history and basic principles of food-production? You must be the judge.

If you heard of someone who had jumped out of a window without getting hurt, you would deduce either that the

1. It was coined in Glasgow by Professor W. J. M. Rankine, 'a Scot of Scots', and was published by him in 1868.

window was near the ground or that a net or some other device had been used. You would not assume that the ability to set aside the force of gravity had somehow been discovered.

I suggest that you may like to look at the history of production of food in a similarly unmystical and moderate spirit. More: I suggest that now you are enabled to begin an undistracted enquiry for yourself.

CHAPTER EIGHTEEN

Lawes; and the Artificiality of Agriculture

Agricultural plants, which practice has shown to differ widely from each other in their respective relations to soil, climate, manuring, and position in rotation, possess at the same time widely differing powers of reliance upon the atmosphere for the constituents which it is known to supply in greater or less degree. – John Bennet Lawes, 'On Agricultural Chemistry', *J. Roy. Agric. Soc. England*, 1847, Vol. 8, pp. 226–60. [This article is also the source of other references to Lawes in this Chapter.]

THE Industrial Revolution was not at first based on steam-power, but rather upon a collection of workers in factories in which the power might be provided by water or by foot. Thus an old survey refers to the poverty of a part of Coventry which was the place 'where the manufacturers live'. In view of the superior amenities of those suburbs and towns chosen by modern manufacturers and company directors for living in, this phrase strikes oddly until we remember that the word *manufacturer* was once meant in its literal sense.

Nevertheless, the Industrial or Mechanical Revolution in so far as it came to be powered by steam may be said to date from about 1768, when James Watt in Glasgow invented the separate condenser for steam-engines and thus made steam power practical for factory use.

About the same time as the Industrial Revolution took shape, there also emerged in north-western Europe what historians tend to call *the* Agricultural Revolution. I question the appropriateness of the definite article, because there must have been agricultural revolutions earlier: as when the Mediterranean peoples took up the cultivation of leguminous food plants and other crops from the East. In the present century there has been another agricultural revolution with a basis and result similar to that of the first.

The agricultural revolution of this century, and the one which began a couple of centuries ago, were both based on deliberate and more or less intensive cultivation of leguminous plants. The Second Agricultural Revolution (this century's) has quite failed to attract the notice of historians; nor has the essence of the First been completely grasped by them. The role of the short-lived red clover and annual beans of the First Revolution – which depended upon taking legumes as arable crops – has not gone unnoticed; but it has been overshadowed by the turnips which also formed part of the rotation, and by the enclosures and other social changes without which rotational farming would hardly have been possible. Yet the turnips alone would not have kept the animals alive in winter; nor could those animals have been fattened without clover hay, or beans, to balance the extra carbohydrate; nor could the turnips and cereals have been grown without legumes and without the virtually new phenomenon of having animals to make dung throughout the winter. The Second Agricultural Revolution was marked by a great extension of cultivation of leguminous hay and seed crops (arable crops, that is), especially in the newer countries; but its most distinctive mark was cultivation of grassland legumes – wild white clover in climates suited to it, and corresponding pasture legumes in slightly warmer climates.

Whether the respective and respected pioneers of these two revolutions knew it or not, they were harnessing soil bacteria. Whether their vision was clear or not – and this is not the place to assess degrees of empiricism – the result of each of the Agricultural Revolutions was to enable more livestock to be kept than could be maintained before their practices were established. Each of the Agricultural Revolutions brought about a big jump in the amounts of first-class protein available for man.

Agriculture is an artificial business. I have small patience with those *laudatores temporis acti* who talk in the strain of agriculture and gardening being close to nature. (Some of these people object to planting of forests in Britain as being unnatural.) Nature never meant large areas of ground to be covered with one kind of annual plant grown in rows: she shows her objection by striving to populate cultivated lands with weeds and shrubs. The first operation in agri-culture is to remove the trees (from an originally forested region like Britain) or to plough up the prairie where grass is the natural cover. Such operations do much more than remove the natural cover; they connote a far-reaching disturbance of equilibria not always reckoned with. The artificiality of agriculture is shown at the start. It is always an interference with nature.

Agriculture need not be directed towards production of food. In as much as agriculture is devoted to food-production, the definition put out by Lawes in 1847 remains good. A slight paraphrase of that definition is: 'Agriculture is the series of processes whereby a given area of land is artificially induced to yield food for more animals and people than it would naturally support.'

The First Agricultural Revolution did not get inspiration or early impetus from use of fertilizers, for they were not introduced until about 1840 – several decades after that Revolution started. Nor, when fertilizers first came in, did they have much effect upon production. The First Agricultural Revolution had a long gestation; it may be said to have become established in western Europe by 1850 or a little later. All that time there was no potassium salt in quantity to give what we now know to be the necessary balance for the nitrogen in guano and by-products of the new gasworks. Use of fertilizers did not become common and rational until about 1890, and then substantially only in western Europe.

The great benefits of the First Agricultural Revolution are mostly to be ascribed to the effects of enclosure together with the greater attention to details of farming which enclosure made possible. All this, of course, in addition to the fundamentals of cultivation of clover and turnips and cereals in rotation, and to the improvements in breeds of animals and in varieties of crops which were continually being made. How the sum of all this appeared to an unusually intelligent contemporary of the later stages of the first Agricultural Revolution may be instanced by a couple of quotations from Lawes:

> The supposition that the artificial manures at present [1846] at our command, might, if directly applied to the growth of corn, be adequate to its sufficient production throughout the county, without the aid of green crops in feeding [animals], is satisfactorily met by such calculations as the following: The county of Norfolk is said to comprise 1,338,880 acres of land: suppose one-half of this to be cultivated on the four-course system, 334,720 acres will be under corn every year. I believe it will not be considered an exaggeration to say that cultivation in this county has increased the natural produce of corn by 10 bushels per acre; and, according to my calculations, it would require something like 50 lb. of ammonia to be supplied in any artificial manure to produce this increase of corn; and considering 1 ton of Peruvian guano to contain 224 lb. of ammonia, it would require an importation of 74,714 tons to supply the necessary amount for one year. This calculation affords some idea of the value of a rotation of crops.

By 'the natural produce' Lawes presumably meant the yield which would be obtained by farming without rotation and without fertilizers or systematic return of dung from animals kept on the new system. By 'cultivation' he meant enclosure and rotation. And:

> In my [Rothamsted] experiments upon wheat, it required 5 lb. of ammonia to produce a bushel of corn. To obtain this amount of ammonia by means of stock, there should be an increase of about 28 lb. of live weight upon the farm; or in round numbers, to obtain 1 ton of grain beyond the natural production of the soil, there ought to be an increase in the weight of stock of 1,000 lb. In order to bring an exhausted soil to the highest state of fertility, it will be necessary to produce an amount of meat by means of imported food (such as hay and oil-cake) as will be equivalent to the increase of grain required. As the green crops increase year by year, the same amount of meat will be produced, but the importation of artificial food will gradually decrease to the point at which the internal and external resources of the farm are so balanced as to secure the largest amount of produce from the soil.

This requires a good deal of annotation if more is not to be read into it than Lawes or anybody else could understand at that time. The amazing percipience of Lawes is here demonstrated very early, since his first experimental crop of wheat was not harvested until August, 1844; and his correct conclusion was in print by 1847. The 'imported food' he mentions would not necessarily come from overseas. By 'green crops' Lawes meant both clover and turnips (as opposed to the two 'white' or cereal crops in the rotation, p. 104). His 1847 point of view may fairly be put as something like this: Rotation with clover and turnips enabled more livestock to be kept; therefore more manure and more nitrogen for the soil; and so on in an ascending spiral until the practicable maximum is reached. But if and while the soil is poor, nitrogen-rich materials (hay and cake) must be imported for a while to give to the animals until the land itself produced enough to feed them well; then the rest follows.

The argument is not complete; yet it is remarkable as it stands. Lawes did not effectively distinguish between the two green crops. Knowledge about the singular ability of the legumes to gather nitrogen from the air was not to come for many years. Lawes was inclined in 1847 (as a young man of about thirty) to credit both clover and turnips with the ability to make use of air-nitrogen. That he was wrong about the turnips reinforces – as we can see – the value of the clover, *without detracting from the correctness and force* of his general argument. It is a plea for balanced and conservative farming maintained at the highest possible level by resort to arable legumes.

The argument is incomplete because it contains no specific reference to possible further gains from grassland legumes and from sustaining the legumes of both cropland and grassland by means of earthy materials. Those gains were to be realized in our century. Now that we have legumes in every possible position on the farm (or at least have the opportunity of putting them in all suitable positions), and possess the knowledge and the earthy materials needed to sustain those legumes in those appropriate positions, there are no more theoretical gaps to be filled.

There remain only the practical gaps left by not making the fullest use of soil bacteria in conjunction with leguminous plants to supply food and to maintain soil condition and fertility.

Lawes's almost uncanny perspicacity was shown on many occasions during his long life. In 1900 he published an observation to the effect that nitrate is the chief carrier of bases for non-leguminous plants. Though he made a fortune out of superphosphate, Lawes's intellectual interest was always in the fate of nitrogen in plant nutrition, and it was most unfortunate that he never succeeded in unlocking the riddle of biological nitrogen-fixation in spite of earnest efforts to do so. He was probably aware, several years before 1900, of the role of nitrate in carrying bases for plants; this phenomenon has important practical consequences, including the effect of offsetting the acidifying effect of nitrogenous fertilizers (page 143).

Although Lawes's observation has been in print for more than fifty years it still seems to be little known, and a number of researchers have been independently making what can only be called stumbling efforts to account for what is sometimes called by the clumsy name of 'physiological alkalinity'. One eminent Californian plant-physiologist lived and died without, apparently, ever knowing of Lawes's conclusion on a point of great interest to him.

In his 1847 paper Lawes mentioned that the action of air in drained land extends to the depth of the drains, and, 'in addition to this advantage, what may be considered as an artificial climate is to some extent obtained' – including a raised temperature of the soil.

One slightly misleading consequence has followed Lawes's thoroughness. In and after 1843 he included potassium salts among the fertilizers he used. This fact seems to have led many people to the conclusion that potassium fertilizers were in general use much earlier than the period about 1890 which I indicate. The fact is that Lawes experimentally anticipated the large-scale commercialization of potassium fertilizers. Even by 1856, when Lawes laid out the first experimental meadow and showed, within two years, the great benefits of stimulating grassland legumes by application of lime, superphosphate, and potassium salts – the commercial sale of potassium salts as fertilizers lay several years ahead. The first factory for production of mined 'potash' was not opened – in Germany – until 1861, and the sale of potassium salts for fertilizer use spread only slowly outside Germany. Still slower was the spread of knowledge about the correct treatment of grassland, though it is now almost a hundred years since Lawes, once again, hit the nail on the head: about the response of legumes to earthy materials.

CHAPTER NINETEEN

A Mad Tea-party

'It's rather curious, you know, this sort of life! I do wonder what *can* have happened to me! When I used to read fairy-tales, I fancied that kind of thing never happened, and now here I am in the middle of one!' – *Alice's Adventures in Wonderland*

I STRESS the recency of the development and attainment of the present level of protein-supply and of all other agricultural and industrial forms of dependence upon fossil resources. An annotated comparative history of the rise of industry and of the congruent expansions and contractions of food imports would be very interesting, and even surprising, if it were presented not in the repelling form of tables of data but in terms of changes in the loci important in agriculture. After all, the populations engaged in industry had to be fed with both food and raw materials at all times; and a brief enquiry into the ways in which this dual demand has been met since, say, 1870 would be, to say the least of it, challenging.

In Britain we tend to think that there is no agricultural history (except what we learnt at school about 'the' Agricultural Revolution which happened so long ago: and possibly something about the advent of tractors and mechanical cultivation). The rest of social history is for us a matter of food imports (about whose origins and distribution in time we have but the vaguest ideas) and of manufactures: this latter rather ruefully linked with the decline of the United Kingdom from the unique position she held last century, and from which she was ousted just sixty years ago.

About the spread of sown clover among the grasslands of Britain since the seed of wild white clover was introduced into commerce by Douglas Alston Gilchrist in 1906 we know nothing: not even farmers do – for every farmer uses the seed to-day, so that its employment has become an unregarded commonplace. Yet wild white clover sowing and cultivation is our British example of that wide contemporary development of legumes almost everywhere during the last forty years or so, which I have called the Second Agricultural Revolution.

The demands upon ores of iron and other metals, and upon combustibles, may be said to have become great only during the last seventy years; they became galloping only within the last twenty. For fifty years or more have people been pondering about natural resources, but usually from points of view related to provision of metals for heavy industry. There still exists a carelessness about fuel. By that, I do not mean to call up evidence of wastage in the usual everyday sense. What I mean is the lack of thought typified by such things as the appeal to synthetic nitrogen fixation and its triumph in relation to the 'inexhaustible nitrogen of the air' – apparently quite losing from sight the draft which that still new process (becoming important about 1930) set up upon not-inexhaustible coal.

The general question of gross reserves of minerals has not yet become acute. There are whisperings of shortages, and there are some bits of evidence about mal-distribution of the mined products, as of other things. However, for world industry as a whole the roof has not yet fallen, and it may not fall in for quite a time yet. But, like a miner threatened with a gradual descent of the roof into the space in which he is working, we can, if we take time off to listen, hear the creaking of the props.

It is only within about the last forty years that civilized man has made himself hostage to his own ability to produce enough effective phosphatic fertilizer by some or any means to continue to sustain – and, if possible, to increase – the new population of leguminous plants which during that short period has been so astonishingly called into leaf and being, to support the greatly increased numbers of men and women and livestock throughout the temperate lands.

How many cattle were there in New Zealand in 1899? How does the production of milk and butter and cheese and meat in 1899 or 1909 in New Zealand compare with that of to-day? And what was and is the proportion of (*a*) exports of such from New Zealand, (*b*) imports of such from New Zealand into the United Kingdom? And similarly for Denmark; the United States; Australia; or any other country you choose?

I don't know the figures in detail any more than I know the history of Mr Pott. But, just as I know something about Mr Pott's earths and what they mean for us all, I do know that whatever answers you find to your enquiries about increases in animal production must be mostly credited to new cultivation, this century, of leguminous plants, which is to say, turning microbiological fixation of nitrogen to account, which is to say plant-available phosphate, which is to say, use of sulphur or other fuel for fertilizer purposes.

From this dependence there is no escape in sight. That knowledge must be squarely faced.

The escape-prescriptions commonly proposed offer no real solution. They usually assume something taken for granted, such as fertilizers coming out of thin air (either literally, as when appeal is made to more nitrogen; or by assuming that the plant nutrients will *somehow* be provided). More irrigation, growing of higher-yielding varieties of crops, breeding better animals, use of hydro-electric or other power, cultivation of microscopic algae or harvesting plankton; if these are not delusory or incapable of speedy realization, they each amount in the end to a requirement for more fuel or fertilizer, and especially for phosphatic and nitrogenous fertilizer which cannot be made without coal or fuel of some kind.

The 'higher-producing animals' must be fed more food, and that extra food must be in good nutritional balance, if they are to yield more. To replace numbers of inferior animals with fewer but better ones may be theoretically

sound, but demands a higher degree of skill from the
feeder and keeper, and probably a higher quality of rations
i.e., less coarse fodder and a higher proportion of protein
Higher-yielding crops will not yield more than inferio
varieties, unless grown on soil that is naturally good or i
artificially maintained in good condition in all respects. An
so on, all down the list.

There is not much hydro-electric power actual or poten
tial. Neither it nor tomic power can produce anythin
material by any means in sight. It is relevant that th
largest steam-powered electrical generating station in th
United States is in Niagara City; that is not the only stean
station in the area; and it is clear that the proximity of th
Falls has attracted consumers of electrical power, of whicl
only about half, however, comes from the energy of fallin
water.

Of a certain proprietary preparation claimed to be o
value in agriculture, an enthusiast is said to have remarke
that it was 'going to put F.A.O. out of business'. A lot ha
been done by the Food and Agriculture Organization of th
United Nations. International action has, for example
instituted a wide enquiry into foot-and-mouth disease in
Europe; has done a good deal towards providing extra pro
tein for Eastern peoples by showing them how to cultivate
tropical edible fish in ponds and other inland sites; and has
taken steps to encourage the growing of lemons in Yugo-
slavia.

Neither F.A.O. nor any other body has notably discussed
the problems of fuel supplies in relation to agriculture. The
'international' view of crop-production seems to be that the
main lines of attack must be education in use and treatment
of better varieties of crops and improved breeds of animals,
coupled with eradication of disease, and intensification of
the use of fertilizers much beyond present consumption
How the additional supplies of fertilizers are to be assured
I have not learnt.

It appears to me that an organization devoted to food and

agriculture which does not occupy itself seriously and solidly with the problems associated with fertilizer supply, to match its other recommendations, can scarcely be said to be in business.

Whenever fuel-for-fertilizer is mentioned, the 'practical' man's objection readily crops up: that, after all, the amount of fuel consumed to make fertilizers is so small in comparison with the other uses of fuel that nobody need trouble about it; the problem, if there is one, can be left to solve itself. So it can, for a short while and on a strictly *laisser-faire* basis. But if expectations of increase in the supply of food are taken seriously, the magnitude of the associated problems about fuels and earths becomes apparent.

As was hinted in Chapter Fifteen, the annual consumption of sulphur – mainly for fertilizers – in Australia and New Zealand together is about 300,000 tons. The population of the Indian Peninsula is more than thirty times that of Australasia. Suppose it were planned to provide the Peninsula with fertilizer (not necessarily superphosphate) to enable it to grow higher-yielding varieties or to adopt other of the prescriptions for increasing its output of food. To be realistic, the Plan must concern itself with fuel, though not necessarily with sulphur, and not necessarily on the present-day Australian scale. Suppose only two million tons a year of coal is allocated to produce extra food in India and Pakistan; just for a start.

Can you imagine the extra shipping, docks, and railway equipment – to say nothing of the factories themselves – which would have to be provided? And would the economic problems be squarely faced?

Such considerations suggest that the relationship of man to the soil bacteria is no longer of the detached kind which, a few decades ago, could be discussed in expositions written under some such title as: *Bacteria: Foes or friends?*

A typical little book of that sort had its first two-thirds devoted to more or less glamorous descriptions of 'man's conquest of disease', and the rest was filled in with accounts of the work of microbes contrasted as 'useful', such as those which make vinegar or cheese. Some of the authors gave a few glances at purification of water and sewage; others went so far as to include references to nitrifying bacteria and to legume nodule bacteria. I fancy that references to actions of the bacteria concerned in the nitrogen cycle were commoner in the popular literature of about fifty years ago than they have been lately: the subject seems to have slipped out of fashion.

Treatments of such quiet subjects were necessarily felt to be dull, because they were remote from human interest and experience. Nowadays, anyone who wants to patronize 'our microbial friends' has his quiver full of antibiotics; though those substances have just about the same relation to soil microbiology and the mechanisms of living as the making of china door-knobs has to the science of geology.

Relationships now existing between man and the soil microbes are too important to be dealt with by dilettantes. Without understanding what he has done or how it came about that he did it, civilized man has placed his future comfort – perhaps even his peace – in pawn to the soil bacteria.

In essence, that is no new thing; for all life comes from the soil, and the products of life are the raw materials of the soil microbes. What is new about the modern situation is that man has cultivated the soil microbes with lime and phosphate and other devices in almost every region, and certainly in every country, where he calls himself advanced or civilized in a technical sense. It is not very material that all this has been done with little thought about the microbes themselves, and that man has usually believed that his efforts have been directed to cultivating plants, not microbes.

However the situation is looked at, it has resulted in propping up an acreage of leguminous plants far greater than the earth was ever called upon to support. That is no bad thing; there is no point in resting the soil from a kindly and conservative system of cropping (meaning one which assures the soil microbes their due place) or in letting it lie idle (unless to recuperate after abuse). Intensive cropping would be a tolerable thing, and could be expanded still further, so long as fossil fuel in sufficient quantity is made available at the right places.

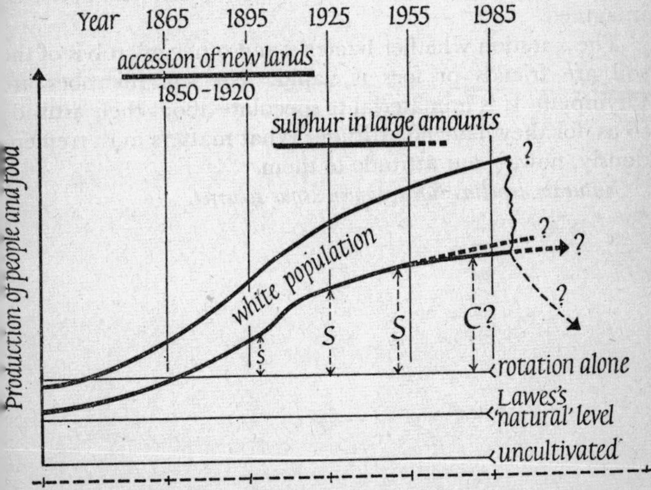

Fig. 5. Diagrammatic representation of the fact that the white population has been increasingly supported by continuous burning of sulphur to make (principally phosphatic) fertilizers. Some alternatives for the not very distant future are indicated.

The lengthening 'props' made of fuel are S, sulphur; C, coal (or other carbonaceous fuel).

Man redeems his comfort, and perhaps his peace, by a yearly contribution of sulphur. He could also do so in the equivalent currency of coal (diamonds for gold), provided that he thinks rapidly enough and to the point.

Otherwise, the amounts of biologically-fixed nitrogen, which is to say, of protein and other foods, may sink back to the amounts on hand in, say, 1899; but there is nothing except effort and ingenuity to stop their shrinking still further, back towards what they were in 1799 or earlier still, when the production of food was in precarious equilibrium with what could be got out of the soil itself and from wind and water. The consequences of neglecting to continue to propitiate the soil microbes at least to the extent we have learnt to render unto them burnt offerings (of brimstone chiefly) do not require to be stated here; they can better be imagined.

The question whether bacteria and other microbes of the soil are friends or foes is vapid. The soil microbes are Olympian. It is immaterial to speculate about their attitude to us, for they have no attitude. What matters most tremendously, now, is our attitude to them.

Naturam expellas furca, tamen usque recurret.

A MISCELLANY OF SIX OR SEVEN

MISCELLANY ONE

The Word *Microbe*

La conclusion de cette étude, c'est donc que la vie ne saurait exister sur la terre, s'il n'y avait pas en même temps, non seulement la mort, mais la dissolution, qui en est la conséquence; c'est que des êtres immortels auraient bientôt épuisé cette source de toute existence, qui est l'air dont nous sommes environnés; c'est que ces infiniment petits, qui ne s'offraient, il y a quelques années encore, aux regards du savant que comme une preuve curieuse de l'exubérante fécondité de la nature, jouent au contraire un rôle immense dans l'univers, et que la disparition des espèces microscopiques entraînerait toutes les autres dans une ruine commune.
– From the summary by M. Danicourt of a lecture given at the Sorbonne by Louis Pasteur. See *Revue des cours scientifiques* [de la Sorbonne], 1864–5, pp. 199–202; or *Œuvres de Pasteur*, Vol. 2, pp. 648–53 [Paris, 1922: Masson].

PERTINENT to our main theme, though incidental to the immediate subject of this Section, is the occurrence, in the same lecture, of the sentence: '*Vivre, c'est en quelque sorte soustraire des gaz à l'atmosphère et les organiser en substances solides et liquides.*' That is to say that life is largely a process of abstracting gases from the air and building them up into solid and liquid substances, or tissues. Throughout Pasteur's early work runs this preoccupation about the connexion between chemistry and life; and indeed it never left him. 'L'œuvre de Pasteur est tout unité' wrote Pasteur Vallery-Radot as one of his telling expressions of the single purpose which animated Pasteur's scientific work and led him to enunciate nearly every guiding principle of microbiology.

The finest of Louis Pasteur's sayings is that which extended the eighteenth-century ideas of the great French chemist Lavoisier about the cycle of life. Pasteur pointed out that without microbes (as we should say now) and the work of microbes in decomposing once-living matter and returning it to circulation, the earth would soon be cumbered with the dead bodies of plants and animals. That idea has occurred to others besides Pasteur; yet none but he has rounded it off with the magnificent sentence to the effect that, but for the work of the microbes, death itself would be incomplete.

These ideas all took shape before the word *microbe* was invented. The quotation given at the head of this chapter exemplifies the awkwardness of such phrases as 'small microscopic species', which had to be used as general terms before the word *microbe* was coined in 1878 by the retired surgeon, Charles Sédillot.

Sédillot was then more than seventy years old. During the Franco-Prussian War he had come out of retirement in order to do what he could for the wounded, and had been horrified and appalled by the unnecessary suffering and losses of limbs and life entailed by the lack of precautions about transmission of infection during treatment of open wounds. Stimulated by Pasteur's ideas about microbes and infection, Joseph Lister had introduced antiseptic treatment into surgical wards in Glasgow and Edinburgh; but Lister's ideas, or Pasteur's, were only slowly adopted in French surgical practice – nor was the French army foremost in adopting them in France. In 1878 Sédillot devoted a lecture to 'The influence of M. Pasteur's discoveries on the progress of surgery'; and it was in this lecture (*Comptes rendus Acad. Sci. Paris*, 1878, **86**, pp. 634–40) that the word *microbe* was first heard in public.

Believing that the word meant literally a being with a short life, rather than a small living thing, Sédillot (who must have had a strict classical education) had serious doubt about the propriety of the word; but Littré reassured him. Thus emboldened, Sédillot said, almost at the outset of his lecture:

> The names of these organisms are very numerous; they ought to be made more definite, and some of them require reform. The word *microbe* having the advantage of being shorter, and also

possessing a wider meaning; and my illustrious friend M. Littré (the most competent linguist of France) having approved of it, we adopt it, though without giving up any of the names commonly used. . . .

The list of synonyms for 'these organisms' given by Sédillot in a five-line footnote is worth reproducing in full. It is: 'microzoaires, microphytes, aérobies, anaérobies, microgermes, micrococci, microzymas, bactéries, bactéridies, vibrions, microdermes, conferves, ferments, monades, animalcules, corpuscules, torules, *penicillium*, *aspergillus*, infusoires, *leptothrix*, *leptothricum*, spores de l'achorium, de favus, de l'oïdium, du muguet, organismes de l'acide tartrique droit et gauche, zymases septiques et septicémiques, etc.'

The word *microdermes* seems to be an error for mycoderme(s). *Mycoderma aceti* is an old name for *la fleur du vin*: a mass of growth of a bacterium able to ferment wine or dilute alcohol into vinegar, and now classed as *Acetobacter* and *Acetomonas* spp.

By 1878 Pasteur had become much occupied with the germ theory of disease. He was glad to be able to give up the use of such terms as 'infusorial animalcules' in favour of the new and comprehensive coinage, so in the same volume of *Comptes rendus*, pp. 1037–43, we find this sentence of Pasteur's:

> Pour affirmer expérimentalement qu'un organisme microscopique est réellement agent de maladie et de contagion, je ne vois d'autre moyen, dans l'état actuel de la science, que de soumettre le *microbe* (nouvelle et heureuse expression proposée par M. Sédillot) à la méthode des cultures successives en dehors de l'économie.

The word was launched; yet, in 1938, when I was writing *Microbes by the Million* for Penguin Books, there were a few heads shaken over the word. Some (like Littré) approved of it (though it does not appear in Littré's great Dictionary of 1889); but for others, sixty years after it had been presented at the court of the Academy of Sciences by irreproachable sponsors, *microbe* was not quite respectable: at least, in English.

MISCELLANY TWO

Two Italian Pioneers

1. HONOURABLE EXCHANGE

THE subject of base exchange has been introduced earlier, with some regret, since it is not one which can be expected to interest the general reader very much. It is a chemical matter, and its best-known, if not the most appealing, exemplification is a method of water-softening. Commercial water-softening by base exchange operates by exchanging the calcium and magnesium in the salts which make water 'hard', for sodium from common salt, through the intermediary of an insoluble proprietary material, which, like the clay and other colloids of soil, possesses the property of holding exchangeable bases and being able in suitable circumstances to exchange one for another. I am not sure whether this brief explanation will be satisfactory either to those who do or did not know much about base exchange.

Base exchange also enters into brewing, because the fibrous substances of malt share with many similar plant materials an ability to bring about exchanges with water (which is always hard in brewing); the malt thus modifies the properties of the water before brewing begins.

There are many reasons for the industrial importance of base exchange. The phenomenon – or rather, an effect of it – was first noticed in 1845 by an enquiring Yorkshire landowner, H. S. Thompson. He wrote:[1] 'As I was aware that the soil had a certain power of absorbing ammonia, I was anxious to test the extent of this power.' Might not ammonia contained in farmyard manure, or in ammonium

1. *J. Roy. Agric. Soc. England*, 1850, **11**, pp. 68–74. The same volume contains Way's famous paper on the power of soils to absorb manure, pp. 313–79; also an interesting paper on warping, pp. 93–113, by T. J. Herapath.

sulphate, be washed out when those materials were added to soil?

Having added a solution of ammonium sulphate to soil, Thompson tried washing it out by pouring more water on. He found that the drainings from the soil contained substantially only calcium sulphate. The ammonia stayed behind, having been left attached to the soil after an exchange with the soil's calcium. Thompson's results were confirmed and extended by another Englishman, J. T. Way, a professional chemist who is often regarded as the actual discoverer of base exchange.

Some Italians claim the honour of discovering base exchange for their countryman Giuseppe Gàzzeri, who from Florence in 1819 published a book on manures.[1] This book mentioned that soil had a marked power of absorbing colouring matter from dirty water and the like. Some Italian friends have been a little peeved by my unwillingness to admit this as a demonstration of base exchange. I had some difficulty in making it plain that this refusal did not arise out of reluctance to cede priority widely accorded to the English; and possibly they are of the same opinion still.

I can, however, claim a distinction shared by few Italians; I have seen Gàzzeri's book. This was not an easy matter to do, because a copy of it is rather difficult to find; so I cordially acknowledge the courtesy of several librarians in Florence to a stranger who happened in on them on a day or two in 1950, and to Signor Ettore Danielli, a young Florentine of high ability and wide interests who was so kind as to prepare the way for me. The book supplied no fresh evidence to lead me to give Gàzzeri any credit in the matter of base exchange; but I found a microbiological feat of some importance – about which no Italian or other writer has yet made a claim, so far as I know, for Gàzzeri as pioneer.

Gàzzeri made an experiment of which the book gives

1. 'Degl' Ingrassi e del più utile e più ragionevole impiego di essi nell' agricoltura.' Firenze nella Stamperia Piatti. 1819. 103 pp. (privately issued.)

details. This related to the decomposition of beeswax and resin in pots of sand planted with wheat. Until recent years, waxy and resinous substances were supposed to be almost unattackable by bacteria. Gàzzeri did not produce the bacteria, or any evidence that microbes were responsible. He was inclined to impute the decomposition to some action of the wheat roots, but he drew the sound conclusion that if such 'difficult' and refractory materials as he experimented with were made so soluble as to disappear, then under the influence of air and water the solid matters of organic manures would likewise become soluble, and assimilable by plants.

He made a very nice quantitative job of showing that the substances he used disappeared fairly rapidly from the sand with which they had been mixed. I hardly know whether to give greater praise to Gàzzeri's experimental technique, or to the discovery itself. Gàzzeri's claim to be a pioneer by many decades in the discovery of the action of the mixed micro-population of well-aerated soil upon wax and resin seems to be incontestable.

II. THE MIRACULOUS BACILLUS: A CORRECTION

IN *Microbes by the Million*, the Pelican forerunner of this book, there was a historical mistake which nobody has pointed out to me. The section entitled 'The Miraculous Bacillus' gave an account of some of the work of Bartolomeo Bizio, a North Italian microbiologist and chemist of some distinction, who flourished towards the middle of last century. The 'miraculous bacillus' is a species of bacterium which produces blood-red aggregations or colonies; appearance of such red spots of growth of the organism on wafers kept in damp churches led to its being given, rightly or wrongly, the English name 'miraculous' and to its being long known under the species name *prodigiosus*. This organism had been studied by Bizio about 1819 and he was the first to give it a name: *Serratia marcescens*. (That name has lately been revived in place of *Bacillus prodigiosus*.) Bizio invented the generic name *Serratia* in honour, as he remarked and as I quoted, of his countryman Serafino Serrati, 'the first who plied a steam-boat on the Arno'. This (one infers) provided Bizio with a little satisfaction that an Italian had put a fast one over the Americans or English (or Scots, if he had recognized the distinction) in the matter of priority of use of steam power for navigation.

This was a sad affair – for me, in one sense, as I now try to make amends – but also for Bizio. There was no truth in Bizio's allegation about Serrati. Serrati was born in Florence, and invented or suggested a number of things, including an oven and at least one contribution to aeronautics; but he worked in the monastery of Cassino, and there is no evidence that he sailed anything mechanical on the Arno. It seems as certain as a historical negative can be, that Serrati had nothing to do with steam power. On the strength of what evidence I know not – but probably as an echo of Bizio's writings, copied from book to book – Serrati has been accorded passing mention in some histories of steam navigation. I am sorry that I have helped to perpetuate that error.

Bizio's attribution to Serrati of a share in the introduction of steam power can only be put down to imaginative enthusiasm, perhaps tinctured with national pride. Serrati's own book[1] described the oven and a means of sailing through the air, but the Letter about ship-propulsion does not mention the word steam, or anything which one would think could be confused with it by an ordinarily careful reader. Yet Bizio went off the rails about it; so did a French encyclopaedia.

There are a lot of bibliographical puzzles in the Serrati business. Several are provided by his own book of Letters. Who edited and brought out the book? To whom were the Letters addressed, and in what circumstances? Was the book published while Serrati was alive?

1. 'Lettere di Fisica sperimentale di D. Serafino Serrati, Monaco Cassinense della Badia di Firenze.' Firenze, MDCCLXXXVII – Per Gaetano Cambiagi Stamp. Grand. – Co Licenza de' Superiori. 81 pp.

The French encyclopaedia provided a more unusual kin
of puzzle. I have not seen the original French version. Th
Museum of Arts and Sciences on the bank of the Arno i
Florence did not have Serrati's book; nor, when I callec
was there anyone who knew anything about Serrati or hi
work; among the Museum's rich collection of scientific an
technical curios there was no model of a steam-boat or any
thing attributable to Serrati. The Librarian kindly looke
up an Italian version of the encyclopaedia[1] and found ther
a suggestion that Serrati was alive when his book was pub
lished. She also found a quite extensive mention of Serrat
under an entry relating to Jouffroy d'Abbans, a Frenchma
to me quite obscure who had some claim to having put
steam-boat on the Saône.

This article reproduced what I later found to be prac
tically the whole of the letterpress of the 'boat' chapter (
can *not* call it the relevant chapter!) of Serrati's book, and i
such a good transcript (there were only a few difference
which could have arisen in copying) that it could not hav
been a re-translation. The French contributor must hav
had the Italian original before him, while he was comparin
the achievements of his countryman with that of the Italian
and waxing hot in making the comparison. Could the un
substantial claims of this wretched Italian to be the first t
invent a steam-boat be legitimately put before or besid
what d'Abbans had done? and so on.

The author of that article was at pains to argue the ques
tion of priority about application of steam to navigation. I
he had read the Letter which he went to the trouble of re
publishing, he would have saved himself some indignation
because there was nothing to argue about. D'Abbans coulc
have been left with all honours due to him; Serrati shoulc
not have been brought into the case at all.

Even in Florence there appears to be now only one copy
of Serrati's book. It is in the National Library, which
closed for a month's stocktaking a couple of days before I

1. Dizionario Biografico Universale (prima versione dal francese),
vol. 3 (about 1845); vol. 5 (1849).

rived. However, the resourceful Mr Danielli found that
hilst a book could not be consulted in the closed library, a
ook from it could be transferred to another library; and
rough such a transfer Serrati's book came into my hands.
ly interest in all this arose solely from the distant connexion
Serrati with bacteriology; but having taken the first step
ong that byway of history I had to continue to its sur-
ising end.

Serrati mentioned a pond as the scene of the trial of his
ttle model. It is to be conjectured that the pond was at or
ear the Cassino Monastery. His small craft had no steam;
or did any under-water machinery break the lines of the
ull. With that model (though not for the same purpose)
errati was doing in miniature what a few years ago was
one on the River Clyde with the old pleasure-steamer *Lucy
shton* after she had had her paddles removed and had been
quipped, for experimental purposes, with pure jet engines
ounted on her top deck. Whether Serrati was the very first
his line I am not able to decide; but he was a pioneer in
t propulsion.

His model had no other motive power than a burning
orch. It was a fire-boat, and by the power of Fire it sailed
ithout wind.

The quaint wood-cut in Serrati's book does not appear to
ally exactly with the text. I think the explanation may be
at Serrati did not supervise the drawing of the illustration,
nd the engraver could not quite understand what was
anted. There should be, apparently, a projection back-
ards; and the cross-hatching on the outlet that is shown is
robably just the engraver's filling-in, and is not meant to
epresent a grid. I must leave such matters to be settled by
ngineers and designers.

The moral of the confusion created by Bizio is, I suppose,
hat one should always consult the original. Bizio's recourse
n this instance to imagination and enthusiasm, in place of
inding out what Serrati did or wrote, was the cause of all
he trouble. I bear him no personal grudge, for the hunt was
un while it lasted; but his unhistorical carelessness, and

seeming wish to score off the foreigner, had the effect obscuring a real merit of Serrati, and of depriving h country for a while of a place in the histories of a novel for of propulsion.

What are we to say about the Frenchman who wrote th encyclopaedia article, and not only had the text of Serrat Letter in front of him, but published it for all historians read? I suggest that the oversight here was understandabl if we recall that to hear or read of a fire-boat could mea only one thing to people of a hundred years ago. If the com mentators of that time thought of a *Comet*, they would have mental image of a smoky vessel on the Clyde: not of a cra cleaving the air between London and Johannesburg.

For my erroneous allusion in *Microbes by the Million* Serrati as a pioneer in steam transport I relied upon a excerpt of part of one of Bizio's publications that was trans lated in a paper by C. P. Merlino in the *Journal of Bacterio logy*.[1] In the limited time at my disposal I was unable eith in Florence or Venice to find this particular work of Bizio though I read something about his researches on Tyria purple and other subjects. Later, the indefatigable M Danielli ran this scarce publication to earth, and sent me partial transcript, which showed that the American tran lation was accurate.

We must blame Bizio; and so, I feel, must Italians, fo having deprived Italy of an honour. When I set out on th trail of Serrati, I was completely unknown to the variou Italians who were so kind to me without having an idea o my purposes; nor did I suspect what the outcome might be so the story ends well, and, I hope, happily for all.

There must remain nevertheless for a time the curiou situation of Gàzzeri and Serrati in history. Each has bee given credit for what he did not do; Gàzzeri for work o base exchange, and Serrati for sailing an imaginary steam boat. Neither has had credit for being a pioneer in hi respective enterprise: Gàzzeri for a good piece of work in field of microbiology not seriously explored until about

1. 1924, **9**, p. 527.

century later; and Serrati as a pioneer in jet propulsion, for whom the interval between the work and its recognition has been even longer.

Nor (it seems) do the twists end there; a microbiological enquiry having led to a finding in the history of mechanical invention, and a chemical enquiry to a discovery in the literature of microbiology.

* * *

It seems worth while to reprint Serrati's Letter (from pp. 74–6 of his book) about the Fire-boat.

Lettera VIII

che dimostra un barchetto a Fuoco che con la forza di esso cammina senza vento.

Essendo un giorno in campagna, e presso d' una piccola vasca, osservai che per essere la giornata quietissima, l'acqua di questa vasca non si moveva punto. Mi saltò in capo di vedere se vi era modo di movere con l'arte l'aria in guisa, che potesse questa guidare un legno senza che l'aria fosse commossa. O sentite di grazia quello che io mi immaginai.

Feci un piccolo Battello, che vedrete descritto B nella fig. 7; nel mezzo del medesimo vi feci fare una gola piramidale G, la quale voltasse la sua bocca orizzontalmente verso la poppa del Battello BB; al disotto di questa gola posi una fiacco // la accesa F, il fumo della quale, veniva a sortire dalla bocca orizzontale A, e formando così come una aerea colonna di pressione, trovava questa una resistenza nell' aria esterna che gli opponeva, formandoli come un punto di appoggio, ed in forza di questa opposizione della colonna del fumo, che incontra nell' aria, il mio piccolo battello camminava.

Dalla figura di questo Battello a fuoco con detta gola, in un colpo d'occhio voi vedete come deve esser fatta questa gola, procurando di farla di lamina sottile perchè non aggravi il battello.

Io non so se questo scherzoso esperimento potesse riuscire in grande, proporzionando la mole della gola alla fiaccola che gli si sottopone, ed al battello; Dico bensì che // quando ciò potesse riuscire, verrebbe questo ad essere di un gran comodo a chi viaggia per mare, per non essere tediato dalle calme tanto noiose ai viandanti. Mi rassegno sempre ec.

A not very elegant translation of the second paragraph is:

I made a little Boat which you will see in the Figure 7. In the middle of the boat a pyramidal trunk G [stands upright], its opening A being turned horizontally so that it faces the stern. At the bottom of this trunk I put a burning torch F [not clearly, if at all, shown in the cut], of which the smoke comes out of the horizontal opening A, thus making a sort of pressure-column. This encounters resistance from the outside air to which it opposes itself, and makes as it were a point of pressure; and by virtue of the opposition of the column of smoke in meeting the air, my little model went along.

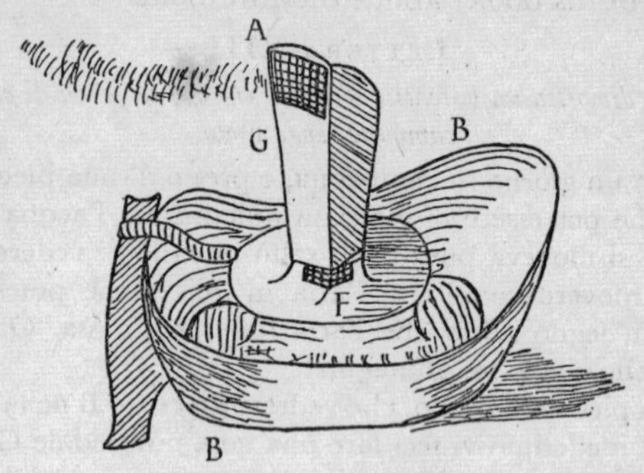

Fig. 6. Serafino Serrati's Figure 7: a copy of a faithful pen-sketch, which, together with a transcript of the text of Serrati's Letter about the fire-boat, were made by Mr Danielli and verified by me.

MISCELLANY THREE

British and U.S. Agriculture

Wisconsin's agriculture has now passed the century mark. Compared to the various agricultural economies of the Orient, the Middle East and Europe, we are still a young and pioneering community....

In Wisconsin, as in all the world, adjustments must be made. Old and new must be sifted and blended to combine the best of the traditional and the modern. – Alfred L. Namejunas, 'Growth of a Dairy State: a profile of Wisconsin agriculture' (*University of Wisconsin Extension Service, Circular 429*, 1952, 11 pp.)

It is refreshing to find one kind of productive effort about which Britons do not have to bow their heads in admiration of United States methods. Little has been heard about a comparison of U.K. and U.S. farming. That may be due to the fact that in Great Britain the majority of people are not interested in agriculture. The publicity given to comparisons of U.K. and U.S. productivity in manufactures may be a reflection of our being preponderatingly urban and manufacturing nations in these islands; also to the fact that there are machines, but no soil, in Fleet Street.

Glasgow is 900 miles north of New York. If the British Isles were slid due west, they would pile up along the coast of Labrador. That the British Isles is not inhabited by a few thousand fishermen and a handful of missionaries is owed to the Gulf Stream. That is a long-standing aid from the Americas; but the blessings which the Stream gives, in making life tolerable in a chancy but seldom bitter climate, are offset by disadvantages. Of these the chief is not actual rain; but rather the high degree of cloudiness born of proximity to the Gulf Stream. Along with that goes a mildness of winter and a coolness in summer which enables us to grow a limited range of crops, but puts beyond reach the possibility of growing maize, or any of the better-known oil-yielding crops, or pulses for human consumption (except garden peas and green beans, in a small area), or – except in a small area – a high-yielding leguminous fodder such as lucerne.

By indicating the serious handicaps under which we live in the United Kingdom my intention is not to exculpate the British farmer but to suggest how our geographical fortunes affect the food of the British public as a whole. Our chief and renowned crop is grass: greatly improved since the introduction of wild white clover by Gilchrist, and a wider use of lime and phosphate and of better husbandry in the last quarter of a century.

Grassland can produce only livestock, which is almost our only domestic source of proteins and fats for human consumption. Most of the rest of our agriculture centres upon livestock. If we were to require important quantities of vegetable protein, or of vegetable oil to use in cooking, they would have to be imported. British advocates of the vegetarian thesis seem to forget that peoples who are vegetarian by custom all live in warmer climates than ours. Vegetarian peoples live where crops directly yielding edible protein and oil grow freely. There is also the point that in such lands the sun warms the air, and thus lessens the demands for protein and oil of any kind. A comparison of the dietetic habits and of the produce of northern and southern Italy is very instructive in this connexion.

The average annual temperature of Glasgow (Renfrew) is about the same as that of Ohio, and is the same as that of Iowa: namely, 47–48°F. (8·5°C.). These are representative States of the Corn Belt or north-central region of the United States. Each State is roughly the size of England; Ohio is a little smaller, is fairly highly industrialized, and has several large manufacturing cities and ports engaged in the iron and steel trades (*e.g.*, Cleveland and Cincinnati – famous for machine tools), and Akron of automobile-tyre fame. Ohio has a population getting on for eight million. Iowa is almost purely agricultural, with no heavy industries and no really

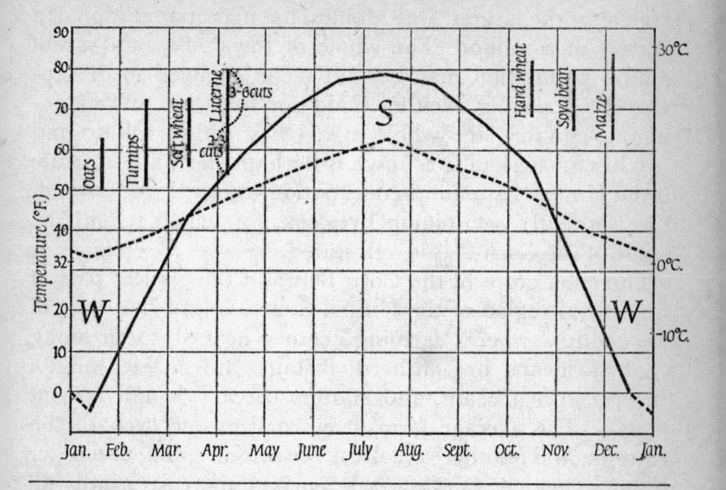

Fig. 7. Schematized diagram to show the approximate course of average temperatures throughout the year in (say) Iowa and lowlands near Glasgow, the mean annual temperature being the same.

The excess of summer temperatures of Iowa over Glasgow is the area marked S. It is this excess which makes possible the growth of maize, lucerne, and soya beans in Iowa, whereas in the west of Scotland even soft wheat is not a sure crop. The fact that winter temperatures are much higher in Britain than in the Corn Belt is of little help for cropping. The areas marked W represent temperatures below the temperature (about 45° F.) at which microbes are active in the soil and crops can grow. Thus Iowa has the effectively longer, as well as a warmer, growing season. Other consequences can be deduced; *e.g.*, the rapid onset of the short Corn Belt spring.

The vertical lines at top left and right show roughly the limits of temperature (during the growing season) which permit growth of various crops. It may be added that some varieties of soya bean and hard wheat ripen in well under 120 days from the date of sowing.

large city; the largest, Des Moines, having a population of a quarter of a million. The whole of Iowa's two-and-a-half million population may be said to be engaged in or supported by farming. Neither Ohio nor Iowa has any mountainous country, the whole area being gently rolling and suitable for agriculture. Iowa is perhaps the 'fattest' State in the Union. Its main preoccupation (apart, I understand, from football) is fattening livestock, for which it and the whole of the Corn Belt is well suited.

The chief crops of the Corn Belt and the greater part of the central region of the United States, *i.e.*, of that part of the country where 'farming' comes nearest to meaning what it means in southern Britain, are maize, alfalfa (lucerne), soya beans, and another cereal – usually wheat or oats. The average farm grows maize, one or both the legumes, and another cereal. In Wisconsin, Ohio, and other States in which dairying is important there is usually an acreage of grass, in which wild white clover is important as in the U.K. In Kentucky – the 'Blue Grass State' – the prominence and indeed glamour conferred upon what the Americans call blue grass (which is the ordinary smooth-stalked meadow grass of Britain, from which it was carried by the early settlers) obscures the equal importance of clover among the grasses. Kentucky is a limestone region, so that supply of lime for its famous pastures is not difficult; however, there, and wherever grass is farmed, a large demand for phosphatic and potassium fertilizers needs to be satisfied also.

The agricultural areas in the United States may be divided into the central and north-eastern regions of mixed farming, including some cultivation of European grasses, and based on rearing or fattening of livestock for meat or dairy purposes; the southern regions, of which one is largely dependent upon cotton and peanuts or ground nuts (another legume!) and the other depends upon sugar-cane and rice; the west-central wheat belt, which is a southerly extension of the Canadian prairies, and which, like those prairies, grows practically nothing but wheat – for the good reason that hardly anything but wheat will grow – year after year; and regions of specialized crops such as the walnuts of California and the citrus of Florida and California; besides the range areas in the Far West, of what is by courtesy called grass.

By way of a change, we might discuss a point about hops, which are not exactly a food crop. In North America these are grown near the north-west Pacific Coast, on either side of the border, but mostly under U.S. auspices. All the hops are concentrated in a few areas of uniform flat land such as occurs nowhere in Britain. This uniformity and concentration go to make mechanization feasible. There is another factor: the hops can be grown very high, so that it becomes worth while to use machines for picking the large bulk of crop presented for a given length of travel.

I have mentioned our cloudiness. Dear British reader, have you ever blamed our curious licensing laws for an alleged unattractiveness of holidays at home? Have you ever wished you could sip a glass of wine on a restaurant terrace at Eastbourne or Ardrossan during a sunny afternoon? There *are* sunny days, when it would be warm, and nice to sit outside, if it were not for the law and the rain (it sometimes rains in France and Italy, even in summer) and the *wind*. If you have been further abroad than Northern France or the North Sea coast, can you recall a single windy day in a summer holiday spent on the Continent?

It is the winds which are prone to chill you when the sun shines in Britain; they whip up the dust, and would disarrange your outdoor table, and would blow down the hops – if those were trained as high as they are in Washington State. English slowness in adopting mechanical hop-picking does not wholly arise from mulish British conservatism; it comes rather from our having hardly any land that is not broken by slopes and swept by strong summer winds. Would *you* care to put up a screen of greenery eighteen feet high, and chance its being blown down (posts, wiring, and all) before you got around to it with a costly machine?

And would you like it if someone who cannot see the

failures of his own country, who cannot see that you have done very well in adapting yourself to what have for a long time been adverse circumstances, and who is inclined to take some credit upon himself for what God has provided in the way of temporarily favourable circumstances, were to criticize you: nominally and superficially for being slow, but in effect for not being lucky or having a large area to choose from to suit yourself?

There are some other points about United States yields. The United States is in many ways conservative – though not, on the whole, with land or other resources. It is the oldest Republic in the world (its formation ante-dates the French Revolution); it is working to the oldest written Constitution (an eighteenth-century one, in essence); and it has the oldest flag in the world (if you ignore the *number* of stars: the pattern has not changed). For measuring yields in terms of capacity it uses the old English bushel, which I think was introduced by Alfred the Great somewhere about the year 900. I may be wrong about the date, but it is indisputable that the U.S. bushel is one-sixth smaller than the Imperial bushel used in Canada and the United Kingdom. This makes recorded U.S. yields look larger. The same five-sixths proportion holds with the gallon and quart; and the common use of a ton of 2,000 lb and the hundredweight of the quite sensible hundred pounds has a similar effect in agriculture and outside it.

In the mixed farming regions the U.S. farmer's problems are agriculturally simple. He is plagued by lack of help, and therefore relies greatly upon mechanization; but our concern for the time being is not with the economics but with the conditions and products of farming. The U.S. farmer's stand-by is maize. This provides him not only with grain for fattening, but with silage for fresh green winter-keep; the preparation of maize silage lends itself well to labour-saving. Thus he is relieved of the great trouble and costs associated with growing the turnips or mangolds which a British farmer must incur, to have succulent fodder for the winter months. With hay made from lucerne or soya-bean crops, the U.S. farmer is well away with a ready-made high-protein fodder such as a British farmer cannot usually have. (Fig. 7.)

The U.S. farmer thus has grain and silage from maize; and probably from his legume hay there is more than enough protein to balance the maize grain and silage. A second cereal – say oats – fills the barn and completes the picture. A British farmer would have to grow five or six crops to do as well, and would probably have to import oil-seed cake or other protein concentrate in addition. He can rely mainly on grassland for dairying or similar purposes, but in any event is likely to be short of concentrated home-produced protein for young stock or really good cows.

It is relevant to note, once more, the recent date of the spread of mixed farming in the main, central region of the United States. For one thing, it has only recently (historically speaking) been settled. For many years after 1805 (see page 253) the interior was unexplored; and farm colonization of the now great agricultural region of the upper Mississippi basin (the Corn Belt, in part) hardly became effective until about 1880. By about 1900 the virtues of alfalfa, and the benefits of fertilizers, were being much talked about. Ten or twenty years later, the gospel of white clover, and phosphate and potassium (and, where necessary, of lime) was preached for grassland – with about the same general success as in the United Kingdom. About 1925 cultivation of the soya-bean (from Manchuria) for hay, oil, and protein began to catch on; with such velocity that within about twenty years the soya-bean became the fourth most important crop of the States: beaten only by maize and wheat and oats, outranking alfalfa, and rivalling the old-established cotton.

Other novelties have been introduced lately into U.S. agriculture. The southern States of the Union possessed in peanuts an old-established legume able to give oil and 'solid' protein; but they lacked a hay legume to correspond with hays from grass or alfalfa. The legumes lespedeza

(*Lespedeza sericata*) and kudzu (*Thunbergia pueraria*) were introduced from southern Japan to help to fill the gap. These introductions were but a small part of the fine work of the United States Department of Agriculture.

More recently still, hybrid maize has come to displace a large proportion of ordinary maize; and the increase in yield per acre has been substantial. Some estimates go as high as 40 per cent. As caution to those (and they are not few) who see in the adoption of better varieties of high-yielding crops a facile and obvious answer to the problem of getting more out of the land, I point out the example of hybrid maize in the U.S. Practically all the publicity attracted, in the U.K. at least, by this innovation has gone to the maize itself.

To describe a new or improved kind of plant is easy meat; its yield can be attractively compared with that of its predecessor; and the article can be lightened with a photograph or two of the crop. It is less easy to bring to notice the cultural changes which have accompanied the growing of the new crop; the photographs cannot show the fertilizer which, on any but the richest sort of land, must have been used – or will soon have to be used – to support the extra growth; and if, at the risk of boring the reader of his enticing article about the marvels of the new crop, the author does put in a reference to cultural techniques, that is likely to be overlooked or soon forgotten in face of the appealing evidence of the photographs. Greater demands are made upon a reader if he is asked to review the historical and technical circumstances which accompany the large-scale growing of a new crop, than if he is just presented with an account of the aspects which happen to lend themselves readily to photography: such as a field of ripe crop and an interview with the breeder.

Though I do not recall having seen a single piece of 'promotion literature' giving advice about cultivation of hybrid maize in the U.S., I am confident that every U.S. resident wanting to buy the seed to try out that crop when it was new had it impressed upon him that if he expected to get higher yields from it than from ordinary maize he must treat the soil accordingly; old methods of just planting the seed and hoping for the best would not do. If the prospective grower was already using fertilizer, he must expect to use more. Such information must have been drummed home by agricultural advisers and also by sellers of the seed – who naturally would not wish their expensively-bred new product to be prejudiced by sowing it under conditions in which it could not show to full advantage.

A reflex of this was seen in Europe when American specialists visited Europe to see what had been done and what could be done to encourage the spread of the new crop. Their findings have been summarized in two reports.[1] From the later of these reports it is difficult not to get the impression that the authors were a little disappointed; partly at finding that there may not be the same broad scope for new varieties in Europe as in the United States, and partly at the difficulty of getting European farmers to use more fertilizer. It would take us too far to discuss whether this latter difficulty is due to conservatism, to economic stringency, or to the fact that fertilizer is no longer a novelty in Europe.

The main point here is that – unlike what so many citizens of Britain, dependent upon imported food and uninstructed in agricultural matters, seem to believe – the American farming population no longer thinks that to buy and sow the seed of an improved variety is enough to get a corresponding increase of crop. For European attention the point is made by the authors of the 1950 report.[2] I give their words:

> It should be emphasized, however, that hybrid maize is not a substitute for soil fertility and that best results are to be expected

1. 'Hybrid Maize (corn) in European countries: Report by United States specialists': O.E.E.C., Paris, 1950.
'Hybrid Maize (corn). Progress in O.E.E.C. countries'; *Technical Assistance Mission No.* 96: O.E.E.C., Paris, 1952.

2. 'Hybrid Maize (corn) in European countries: Report by United States specialists': O.E.E.C., Paris, 1950.

only from hybrids grown on fertile soil. It should further be emphasized that the larger yields from hybrid maize constitute a heavier drain on soil fertility and, therefore, must be offset by suitable soil management practices, including larger applications of plant food materials, if soil fertility is to be maintained.

Dry-land wheat, associated with a belt of States running south from the Canadian border, and thus in part with the 'dust-bowl' so notorious in the thirties, does not for the most part require use of fertilizer. That these formerly prairie soils are naturally rich in lime and the other earthy nutrients of plants is not wholly an accident: it is a consequence of the low total of annual precipitation and other factors which reduce leaching to a minimum. The soils thus came long ago into balance with the semi-arid climate and their natural cover of native grasses. Since being broken up a few decades or years ago, they have lost an appreciable fraction of their organic matter and its stored nitrogenous compounds, but there is still enough of the original nitrogenous organic matter, plus what is formed annually,[1] to permit growth of what by British standards would be judged a low yield of wheat.

Fertilizers are of no use on semi-arid lands, because the limit to production is set by the soil moisture, not by the nutrients with which it is in equilibrium. It seems likely – if no droughts occur like those during the thirties which drastically showed up the chief weakness of this form of extensive mono-culture – that such soils will continue to bear their low per-acre yields of wheat indefinitely – provided they are not irrigated.

Excepting dry-land wheat, therefore, it can be said that all the novelties and expansions of crop with which U.S.

1. Though the free-living nitrogen-fixing bacteria such as *Azotobacter* spp. are not often credited with an important contribution to productivity of soil, it seems probable that they furnish an appreciable part of the nitrogen available to plants in semi-arid soils such as these, so long as the stock of organic matter is fairly high (*i.e.*, on condition that the soil is virgin or that its organic matter still forms a high proportion of what was in the virgin soil).

griculture has bedecked itself during the last fifty years or
) have been accompanied by growth of consumption of
rtilizers. This increase has been very great indeed in the
ıst twenty years. A further big increase is forecast in con-
exion with what is called 'the fifth plate'. That is a phrase
hich means that by about 1975 there are expected to be
ve U.S. citizens for every four alive in 1950.

Excepting the commercialization of wild white clover
ed since 1906, and the later introduction of sugar-beet, we
ı Britain have nothing to match the novelties brought into
merican cropping. It should be remembered that white
lover was wild throughout Britain. Though often difficult
) find, this perennial clover was present in most grassland,
nd could be brought to form an appreciable proportion of
ıe herbage if the right conditions were provided. Gilchrist's
nmediate predecessors found that a system of management
f grassland based on use of basic slag (a fertilizer containing
me and phosphate) encouraged literally wild white clover
ı the bleak Northumbrian pastures to the extent that they
ecame much more productive in terms of animal output;
nd analogous observations about the effects of crushed
one as manure had long before been made on Cheshire
airy farms, and elsewhere.

Gilchrist's introduction of wild white clover seed led to
s becoming a normal ingredient in mixtures intended for
wing-down a grass field. There being insufficient of Ken-
sh 'wild white', the seed used in the U.K. is often that
nported from New Zealand or Denmark or some other
ountry which, like our own, reposes its animal production
pon legumes and phosphate, and can – unlike most of
ritain – ripen the seed of clover. The clovers overseas all
ıme from Europe originally, in the same way as the blue
ass was taken to Kentucky (page 228).

So, from Britain we have in the course of centuries ex-
orted both livestock (as is well known) and – as is much less
ell known – the principal fodder grasses and clovers on
hich those animals, or their descendants, feed; but we have
) instance in the last hundred years of the successful intro-

duction of a new crop such as alfalfa or kudzu or soya bea
from distant countries, or of new varieties comparable wit
hybrid maize. Sugar beet came lately; but from no furth
than Europe; and on our limited area of suitable land
could do no more than displace turnips in eastern Britain.
replaced a 'feed' crop by a 'cash' crop; but there is muc
doubt whether it added anything to nett production, or
except indirectly through raising some standards of cultiv
tion – to soil fertility.

We cling (if that is the word) to the old-fashioned moul
board plough – which suits our conditions – and make litt
use of the more modern disc plough commonly used in t
wide open spaces, where it suits the conditions prevailin
even though the agriculture there is not always what
European would regard as farming. We go on growi
turnips, though their cultivation involves much more labo
than would be needed to grow maize; and so on.

Lucerne or alfalfa (*Medicago sativa*) was taken westwar
by the Spanish Conquerors. Four hundred years ago th
had established it on the west coast of South America, ar
possibly further north. Alfalfa is first known to have be
brought into California by gold-seekers of 1849 who we
round the Horn and thought the seed would be a good thi
to take with them from Chile. From California it was tak
overland eastwards, by people fortunate or disappointed
the Gold Rush; it travelled by slow stages across t
West, Nebraska, and the Plains until by 1890 it had reach
Ohio.

There was another important centre of dissemination.
1859 the immigrant, Wendelin Grimm, had brought a f
pounds of alfalfa seed with him from Germany into Minr
sota. This was a more 'northern' strain, better adapted th
the Chilean kind to survive frosty winters. In the course
years Grimm selected and multiplied a variety still kno
by his name; and by 1890 alfalfa was being talked about,
over the Corn Belt and the East. It had arrived to ga
recognition as a crop with great potentialities.

Alfalfa 'took on' rapidly with the old settlers and with t

ods of immigrants still pouring into the new lands or ying to make a living out of farms recently abandoned, ving borne no legume (because there was practically none the market) and having been cropped with a succession crops such as maize, oats, timothy or native grasses for y.

In 1896 the Department of Agriculture sent out an pedition in search of newer varieties of alfalfa. The S.D.A. has sent many such expeditions all over the world hunt for new varieties of plants and animals and new ethods of combating pests: this was the first of its explorans for a legume.[1]

Some echoes of the fame of the soya-bean (under names e the 'wonder bean') reached Britain from the United ates in the thirties, and attempts were made to grow it re; but not even the quickest-maturing variety could be pended on to ripen even in southern England. The soyaan is a native of Manchuria and northern China. It had en grown in the United States about 1800, but had racted no special notice much before 1925. Then it ught on.

Here are a few U.S. figures, soya-beans first:

In 1907	50,000 acres
1935	5,500,000 acres.
In 1920 the crop produced	3 million bushels (U.S.) of seed;
In 1935 the crop produced	40 million bushels (U.S.) of seed;
In 1946 the crop produced	201 million bushels (U.S.) of seed.

sides which, there was some grown for cutting or harvestg green (not grown on for seed). Between 1936 and 1946 U.S. acreage of soya-beans quadrupled. The crop was imated to be worth 14 million dollars in 1929. In 1944

1. The records of these should make interesting stories of travel and enture. Some account of American and British enterprise in hunting insects in many parts of the world has been given in the author's ican book *Biological Control of Insects* (Penguin Books: 1943, 174 pp.). also *Chambers's Encyclopaedia*, Vol. 3, *art.* ENTOMOLOGY, ECONOMIC ndon, 1950).

the soya-bean *industry* had increased in value to 395 milli
dollars.

Of alfalfa there were over 15 million acres in 1948, t
crop being valued at 250 million dollars. There were s
States each having more than a million acres (1,500 squa
miles) of alfalfa.

In view of the absence in Britain of novel crops a
cropping methods, it was perhaps not surprising that so
Americans should think that – while we were sound
heart – there was something lacking; perhaps zip and pep
our agricultural advisory services, which were not (it mig
seem) doing all that might be done to bring to the notice
British farmers the advances which surely should be possib
if only they copied the eagerness of United States farmers
take up new crops.

Having thought, it seems, on those or similar lines,
influential American was instrumental in getting a numb
of British agricultural advisers to have a look at U.
advisory services to farmers. Before going further I must,
they say in Parliament, 'declare my interest'. I am Ag
cultural Advisory Chemist for the West of Scotland. I thi
however that the aforesaid influential American was
concerned with the agricultural sciences (advisory agric
tural chemical services seem to be poorly developed in t
United States) but with 'putting across' novelties in cr
husbandry and animal husbandry.

Two of my colleagues engaged in what Americans c
'farm extension work' and 'farmer education' went
accordance with this scheme. One of them was an agric
tural adviser; and being an intelligent man, he found mu
that was striking, but hardly anything applicable to Brit
conditions. In Ohio he asked how they would regard t
prospect of feeding forty millions from the produce of t
State, and got a dusty answer. He brought back some col
photographs; and among the brightly-coloured barns a
other features of the American rural scene, one of th
pictures has stuck in my memory. It was of a naviga
river, equipped with spacious and efficient wharves, a

bearing barges of a capacity unmatched over here. The river was brown; looking like a street of liquid mud. I had seen photographs of American rivers often enough, and knew something about soil erosion and its consequences; but I had never seen, or imagined, just what this colour-photography showed.

Among the failings of Americans is a tendency to judge most things by a mechanical standard and to ignore other aspects of civilization. Some delight to point out to the French that the number of tractors per thousand acres in the U.S. is about seven times (or whatever the figure is) the number in France: the inference being, no doubt, that American agriculture is about seven times as good, or as progressive, as French agriculture. Yet an average U.S. farmer, and many U.S. agricultural engineers, could learn a good deal about care and respect for the soil, from an average French peasant. The truth is that British and European agricultures have long been in practical equilibrium with their soils (not necessarily with their populations, of course); agricultural systems having been evolved in Europe which are stable with respect to soil and climate, though in a few geologically recent areas there is unavoidable trouble from erosion of geological type.

The United States, with its short and often catastrophic history of relations of white men to the soil, is still in the formative stage of agriculture. In an area as big as Europe but having a more varied climate, containing the largest agricultural plain in the world, and largely unaffected – for good or ill – by the ideas and practices of white civilization until about a century ago, there is ample scope for development of introduction of new crops and new methods so as to take the best from experience everywhere.

Since we cannot survey every region of the United States, and because some attention has been given to the bulging Corn Belt, and the semi-arid wheat regions, let us look at a less extreme region: Wisconsin. The State of Wisconsin is the premier dairying State. It is slightly smaller than England and Wales combined, and it contains about 3½ million people. Because the market for fresh milk is limited, much of the produce is condensed milk (of high quality and of brands not unknown in Britain) and a wide variety of cheeses. The climate favours grass-legume mixtures, or 'tame hay' (as it is called) of the types based on alfalfa and suitable clovers. Except in the uplands of the extreme north, maize can be grown. Tame hay, oats, and maize occupy 90 per cent of the cultivated land. Besides maize, there are few crops which we should regard as exotic; almost the only speciality is cherries, grown especially on the peninsula projecting into Lake Michigan.

The State resembles the southern part of Britain in having been heavily glaciated except in its furthest south, and in having a northern region unfavourable to agriculture. Clearing of the forest cover of Wisconsin did not begin much before 1830. According to Mr Namejunas, statehood was granted in 1838. In 1860 the State was important as a producer of wheat, the north being almost virgin. Gradually the frontier of cultivation crept northwards into the trees; it took the thirty years up to 1910 before the eleven counties in the 'middle-north' had over half their land in farms.

Expansion of alfalfa acreage, mainstay crop of Wisconsin dairy farming

Between 1927–1936, an average of 514,000 acres of alfalfa was grown in Wisconsin. Dairy farmers of the state faced a feed and forage problem. Research scientists, specialists and Extension service combined forces to increase alfalfa acreage.

To grow alfalfa successfully, Wisconsin soils needed extensive liming. Information and education on soil testing, programs for providing farmers with limestone at low cost, and development of local liming services laid foundations for an expanded alfalfa acreage of 2,000,000 acres in 1951.

Thus Circular 429 (see page 225); though one may suppose that calcium was not the only deficiency which needs to be made good in Wisconsin's leached, recently-forested soils. The last 800,000 of the two million acres of alfalfa were put down in the years after 1948.

This increase of 2,200 square miles under lucerne was a

triumph for the advisory agricultural service. It must have meant a great deal of work and travelling and argument in persuading thousands of stubborn and uninstructed people to do what had already been done by the growers of the first half-million acres of alfalfa. But Wisconsin could grow alfalfa; that must have been evident to the farmers of the next two-thousand-odd square miles of alfalfa, if they had ever travelled more than a few miles; and automobiles and other facilities for communication were not unknown in Wisconsin in the thirties.

So, the real point is not the excellent 'salesmanship' of the Extension service, but the immaturity of much of the farming. My agricultural colleague who visited the United States said that the best farmers there are as good as any in Britain, but there is in the United States a much greater proportion of 'C' or third-class farmers.

There was in Wisconsin in the period under review a protein shortage for livestock; alfalfa could overcome that deficiency; alfalfa could be grown, but was largely not being grown. The same was widely true of the United States earlier, before alfalfa and soya-beans and other legumes were taken up – one might almost say, embraced: and virtually only during this century.

Was the increase of 1,500,000 acres in Wisconsin under the needed protein-supplying crop a credit to the farm advisory service? Undoubtedly, yes. Does a large increase of late years in the acreage under legumes show that U.S. farming is progressive in a 'plus' sense? Or does it suggest that Wisconsin and other States have been catching up with arrears of the possible? (There is a difference!)

In Britain our trouble is of the sort that if we had some eggs we could have some bacon and eggs, if we had the bacon. We cannot hope to compete with the U.S.A. in largeness of figures, but if we could put half a million acres of the Welsh hills under lucerne it would have been done long ago. And if we could do that, we could also grow corn to balance the protein.

Out of nearly 24 million acres classified as farm land,

Wisconsin has about $12\frac{1}{2}$ million acres of cropland. At the end of the second World War – the latest relevant date given by Circular 429 – there were 2,585,000 dairy cows in Wisconsin. There must be more now: say three million. I regret that I am not sufficiently familiar with the American language to say whether 'dairy cows' means what it says (*i.e.*, cows for providing milk for sale, as distinct from cows of beef breeds) or whether it includes bulls, etc.

Little Scotland with a relatively small arable acreage has three-quarters of a million dairy cattle (including young stock, heifers or 'queys', and bulls) and an equal number of beef cattle. England and Wales have about $5\frac{3}{4}$ million dairy cattle and $1\frac{3}{4}$ million beef cattle. The numbers of dairy cows in calf or actually in milk would be for Scotland, and England and Wales, about 240,000 and 2,880,000 respectively: a total of $3\frac{1}{8}$ million. These just about manage to supply 49 million people with all their fresh milk, and a small proportion of their butter, cheese, and condensed milk. In addition, Britain (with Northern Ireland) produces enough other food, on an area smaller than Wisconsin and Ohio combined, to feed about 25 million people.

In the United Kingdom of Great Britain and Northern Ireland we entirely lack those rumbustious success-stories about millions of acres of alfalfa, soya-beans, maize, and other staples developed during the last generation or two. But need we be ashamed? Which twin needs the productivity team?

Let us recognize that in some matters of research and teaching and practice with food the advantage is not so clearly ours. Thus, the University of Wisconsin is among the world's leading centres of research into nodule bacteria and biological nitrogen-fixation; it also has acquired in its short history a very distinguished record of discovery of vitamins and of investigation into other important aspects of nutrition. Nearly every State has a University School of soil bacteriology; in some States there is more than one such centre. In Britain the centres of training in soil microbiology which are as long-established and have as much influence as each of several U.S. and Canadian ones can be counted on one finger.

The reasons why novelties in husbandry catch on in the United States are, firstly, the energy and resourcefulness of the U.S. Department of Agriculture and numerous other bodies concerned with introducing, testing, and preaching innovations – and also in recommending the doing of things which British farmers constantly practise; and secondly the fact that there is so much room for the 'novelties' (new and old) and so much slack to be taken up. There is credit to be given for energy displayed at official levels; and, at the level of actual farming, there is a readiness to adopt the easier ways. All that, and much more that is admirable in U.S. agriculture, does not justify criticism of British or European farming methods as unimaginative; and those who judge European methods of agriculture by as yet unestablished criteria deemed valid in the States show themselves as poor historians and geographers, and as being parochial and not a little blind.

The increases in yields consequent upon the introduction of hybrid maize in the United States were contributed to by wider adoption there of some practices which British or western European farmers largely take for granted (and which therefore do not form an important part of the instructional activities of agricultural advisers here).

Among these old-fashioned, long-established, and helpful practices is rotation of crops. Rotation is axiomatic in European arable farming; in Scottish farming rotation is extended to the grassland.

This contrast of English and Scottish practices supplies a domestic model of an old method having to be preached as a novelty. For some years[1] there has been fairly active propaganda in England and Wales to induce the farmers of those countries to rotate their grassland – 'to take the plough all around the farm' – in the manner long accepted

1. Approximately since the publication of an article by W. A. C. Carr on Scottish farming in *Journal of the Ministry of Agriculture*, 1924, **31**, pp. 14–20.

in Scotland. That excellent and well-balanced little book *British Farming* (see page 246) makes a dry but apt comment on this. It mentions that in England there has been discussion about a suitable name for this all-round rotation some call it 'alternate husbandry', others 'ley farming': 'in Scotland it has no special name – just fermin'.'

Adoption of rotation implies use of a leguminous or 'grass' crop (and the means to sustain the clover or other legume). It is one of the means of conserving soil organic matter and promoting soil fertility.

In the peculiar conditions of the United States, rotation can do more, even, than it can in most parts of Europe; for a crop which completely covers the soil (unlike maize or cotton) is the most powerful mechanical preventive against soil erosion. Rotation – including growing of soil-covering leguminous crops and grass, return of farmyard manure to the soil, and accompanying use of lime and fertilizers where necessary – could thus be of outstanding value to the well-being of the United States. That is recognized by all agricultural authority in the States; yet in many sections of the farming community of the United States the benefits and advantages of an appropriate rotation still have to be preached as hot gospel.

In the States, lack of attention to conservation of soil organic matter and a widespread attitude to the soil as if it were no more than something to be engineered in order to yield tolerable crops, have led to extensive soil erosion. In the moister regions it was manifested by many effects of water-erosion, of which the broad stream of brown mud already mentioned is only a sign. Erosion by wind in the drier regions brought the 'dust-bowl' into European consciousness

It is possible to exaggerate the agricultural effects of soil erosion. In the moister regions, at least, there is probably no result of erosion that cannot be mitigated, or cured, by adopting farming techniques which cover the soil with crops for at least part of a rotation. In that way the effects of hot sun and intensive rainfall upon bare soil in lessening the soil's content of organic matter – and all that that diminution implies – can be exchanged for a proportion of organic matter somewhat nearer that originally in the virgin soil (which was necessarily in balance with the prevailing climatic conditions; not very long ago – historically speaking – all the streams of the United States ran clear). In some circumstances there may be no cure except by grassing the land completely, or afforesting it; and thus going out of arable farming, or out of farming altogether. Without discussing the extreme alternatives, we will keep to consideration of the areas on which general farming can be continued, albeit, it may be presumed, at some level below European standards.

The outline of a rational method of attack has been given. To repeat: it must depend upon encouraging a system of farming whereby organic matter is conserved, and that largely by use of leguminous and grass crops which cover the soil and yield protein for animals. By the aid of such crops will fertility be raised: not merely because of the nitrogen added by the legumes, or because of the extra fertilizers employed in conjunction with the whole rotation. In poor or unimproved soil the inorganic nutrients, if used alone, can do little more than increase yields for a few seasons; it is the organic matter which is the by-product of what we may for brevity call grass-arable farming, or grass farming, that raises soil fertility and improves soil condition.

Everything considered, it comes to look very probable that recent U.S. increases in farming yields per acre or in total are not evidence that United States agriculture is on the whole better, or more up-to-date, than British. The fact that marked increases can be obtained is, rather, evidence of a generally lower standard of farm productiveness: the increases started from a lower level.

I feel sure that in the United States there is still a lot of leeway to be made up before the agricultural ship comes to an even keel upon a rational course. (I make a present of that phrase to the politicians.)

There is little doubt in my mind that the fifth plate will be filled. It can be filled by bringing U.S. farming nearer to

standards which already obtain in old Europe. Still, whe
we have gone as far as is practicable in giving major demon
strations of peoples in equilibrium with the land, it is a littl
annoying to be told that we are the unprogressive ones.

FURTHER READING

Agricultural Extension Services in the U.S.A.: Report of a working party o European experts: O.E.E.C., Paris, 1951, 200 pp.

[Odd fact: from thirteen countries represented in the Organizatio for European Economic Co-operation went seventeen delegates from the United Kingdom went three, of whom one from Scotland Luxembourg, which is the size of Perthshire, sent two.]

American Agriculture: its Background and its Lessons, by A. N. Duckham H.M. Stationery Office, 1952, 78 pp. 2*s*. 6*d*. net.

British Farming: H.M. Stationery Office, 1951, 97 pp. 3*s*. 6*d*. net.

MISCELLANY FOUR

Soil Condition and Conditioners

I bounded across the grass, industriously raving and praying by turns. They were lying on their stomachs and looking over the edge of the cliff. I approached them on tip-toe, threw myself upon the ground, and grasped a foot of each child.

'Oh, Uncle Harry!' screamed Budge in my ear as I dragged him close to me, kissing and shaking him alternately. 'I hunged over more than Toddie did.'

'Well, I – I – I – I – I – I – hunged over a good deal, *any* how,' said Toddie in self-defence. – *Helen's Babies*, by John Habberton

I HAVE hinted at the powerful functions of soil organic matter in maintaining soil fertility. Since 'soil fertility' means different things to different people, and is often used without definite meaning being attached to it; and sometimes in loose or superfluous ways (as when it is taken to be the same thing as annual productivity, and might then be taken as synonymous with 'yield' or 'increase of yield'), I state the meaning I attach to it. By 'soil fertility' I understand the ability of land to maintain a yield of crops for a considerable period. It is thus roughly equivalent to what a good farmer calls, and respects as, 'condition'.

Thus understood, soil fertility is a product of the action of soil microbes and micro-organisms upon organic matter *in equilibrium with* the climatic and other conditions. The equilibrium may be more or less natural, as in the prairie lands of western North America (whether unbroken, or cropped with wheat), or it may be artificial: one to which it has been raised by the efforts of the farmer, as in British practice; or one to which it has been depressed by less conservative methods of agriculture, as in many of the more humid parts of the new countries; where soil fertility has been lowered.

Probably all readers will be aware of the publicity given to synthetic substances as soil-conditioning agents, or 'soil conditioners'. It appears that these substances are in many ways effectual in doing what is claimed for them: as by improving the permeability of soil to water and air, and maintaining soil cohesion so as to improve the structure of soil which has been run-down in condition by inappropriate methods of cropping. Suitable synthetic soil-conditioners will, apparently, be able to do all that the organic matter of soil can do.

That may be true; and this is not the place to examine the justice of the claims for the new soil-conditioning agents. I propose merely to examine briefly their philosophical aspects; some people might, indeed, call them the practical aspects.

The country which has the greatest need for the old-fashioned soil organic matter and other soil-conditioning agents is the United States. That country appears to be the prime market for the new synthetics. The area of run-down land, which might be regarded as badly in need of conditioning, might be taken as a total of 50,000 square miles – roughly the area of England – over a half-dozen average American States concerned with arable farming. Supposing a synthetic conditioner were applied to that area at the commonly suggested rate of one-tenth per cent of the weight of top-soil, *i.e.*, at one ton of conditioner per acre; and that to make one ton of synthetic conditioner it is needful to consume about three tons of fuel (a probably conservative guess, lacking actual figures), then one treatment of the given area of 50,000 square miles with a synthetic conditioner would involve the consumption of about 100,000,000 tons of fuel. That amount is of the order of half the annual output of coal from all British coalfields.

Fifty thousand square miles is a small area in relation to either the world's land surface or the area of farmland. On a globe twelve inches in diameter an area of that size would be represented by a square centimetre; or about a fifth of the area of an ordinary postage stamp.

Such a piece of stamp stuck on the map of, say, North Africa (or North America) would almost escape notice. Yet, to supply one treatment with synthetic soil-conditioning agents to the corresponding actual area would abolish a not inconsiderable fraction of the world's reserves of carbonaceous fuels.

It may not happen. Should it happen, or threaten actually to happen, I should find it difficult to apply a word or a phrase fit for the folly of those who, having within a short time abused and wasted their resources of soil and land, would cap their neglects by making large and fresh, but irreparable, incursions upon resources of fuel.

To that irrationality there is a rational alternative in adoption of conservative methods of farming: meaning by that, agricultural methods which conserve soil organic matter and fertility. Without a potential demand for fuel with which to make synthetic conditioning-agents, there is a real and present demand for a consumption of fuel for making fertilizers and for other purposes connected with food.

For the sake of sanity, let us hope that the worst follies may not be perpetrated. The phenomenon that synthetic soil-conditioners should be more than briefly thought of as a considerable adjunct to agriculture, or as a solution to the larger problems of farming in the newer countries, is just another expression of that heedlessness about fuel of which several examples have been given.

MISCELLANY FIVE

2154

> If the reader ... feels any discomfort over the prospects of all our present fuels being gone in less than 200 years, let him relieve it by a little philosophizing sparkable by a backward look for the same distance. Back there, none of our present fuels were considered of much use or dreamed of. – A correspondent (name withheld) writing from Newark, Del., to *Chemical and Engineering News* (19 October 1953; p. 4282).

THIS is enough for a British chemist to initiate a dissertation about the philosophies of those irrepressible Americans. None of their leaders of thought – with the exception of the authors of the book mentioned on page 163 – seem to have devoted serious attention to what may happen to food supplies if fuel and ore supplies are diminished below the present level. Dr James Conant, speaking in London, remarked that somewhere about the turn of the century the prevalent belief in hell fire appeared to give way to doubts about the duration of fuel supplies on earth; but he seems to have been thinking about the importance of fuel for industry and transport only. An address he gave in the United States about looking 'into the crystal ball', and other utterances by American chemists and technologists, suggest that the general tone is of optimism, born of reliance on fresh exploitation of sources of external energy, and engendered without full reference to the biological facts of life.

If fuel is exhausted by about 2150, it will be easy to show that there has been progress as compared with 1750, when the whole world depended almost exclusively upon annually renewable sources of energy – sulphur for gunpowder being the most general exception.

The population sustained by the food which could be wrung from the earth and seas about 1750 was of course far smaller than it was in 1950. Two hundred years further on, by use of modern methods of cultivation (such as harnessing a mule with ropes to a pointed stake; and the return of all wastes to the soil, as was done immemorially in China, where use of fossil fuel has never been an important feature of agriculture) the present area of the United States will be able to support several millions in the 2150 American way of life. This will indicate considerable progress over 1750, when the total population – red, white, black, and Eskimo – of North America can hardly have exceeded a couple of millions.

* * *

It is 2154. Three hundred and eight years ago, the United States first extended from coast to coast. The time which has elapsed since that event is shorter than the period between its happening and the arrival of Columbus in the Americas.

Travel in the United States is highly organized. There are knowledgeable guides ready to accompany parties of travellers along well-defined walking paths stretching everywhere across the continent. Snakes have been destroyed; and, since there is no danger from Indians, this represents great progress in facility of communication since 1805.* Some plutocrats, able to afford animal feed, go on horseback.

The harbour of Boston has a few wooden sailing-ships, bringing luxuries such as tea, much as in 1770. Bostonians and others recall the tales told by their grandparents, about houses heated by storage of solar energy,† until the system had to be abandoned through shortage of metal.

In the Far West, buffaloes are cultivated on a small scale, using the native range within wooden fences. It is understood that plans for large-scale herding are unlikely to be carried out on account of the difficulty of bringing lumber from the producing regions.

Acetic acid, and a small tonnage of the surplus corn

[maize] grown near the coast, is exported to Eastern nations, in exchange for necessities such as drugs. In China some coal is extracted by hand; this forms the basis of the world's chemical industry, plus products of wood distillation.

There being no chemistry left in North America, there is no tradition of agricultural chemistry. Exchange of corn for drugs is an unsentimental business arrangement. Teaching is largely in the hands of political economists. There survives among them a legend to the effect that about 1950 their predecessors conceived the Columbus Plan[1] for helping the East to produce more food, but without provision for fuel to sustain the scheme.‡ Economic historians cite this alleged Plan as an instance of the inadequacy of old-fashioned thinking to foresee developments; but in justice to their forerunners the moderns must admit that the peoples of the early twentieth century had scarcely had enough opportunity to learn by experience about the relation of fuel-supplies to food-production: a connexion which had been amply demonstrated by history after 2050.

There is still some coal left in the United States, but it lies so deep that there is not enough ore to equip the mines with the elaborate installation needed to get it out. The nation's population is now, therefore, in balance with what the land can produce.

How the population declined from its maximum of two hundred million or so is not exactly known, owing to long prevalence of chaotic conditions before a natural equilibrium was attained.

The greatest drop was after Fuel War II. This had been fought between the nations who badly wanted fuel for their agriculture and fishing-fleets, and the others who wished also to go on burning fuel in their automobiles and civil airplanes. Nowadays, uneducated folk look upon it as queer that people should ever have tried to kill each other over a

1. Some scholars conjecture that the word is Colombo: it refers, they think, to a small island which already about the mid twentieth century was one of the territories having difficulty in feeding its population.

mere matter of equilibration. Others, better informed than the generality, regard the Fuel Wars collectively as the inevitable working-out of the *élan vital* or Life Force, and accept them philosophically. Though we have our troubles even in this year of progress, we count it as one of our modern blessings that we are no longer subject to popular strife over ecological themes.

Unlike the wars of the twentieth century, Fuel War III was relatively expensive in fuel, since the struggle for what was accessible used up much of what was left. Casualties were fairly light in what is called Fuel War IV; though this was not really about fuel, but about the last ores. It was fought mainly in Africa, and was followed by what seems to have been the last great decline in white population.

REFERENCES

* In 1805 Lewis and Clark were the first white men to cross the mainland of the present United States so as to approach the Pacific from the east in that part of the world. In 1793 the Scot, Mackenzie, reached the Pacific by crossing the north-western tip of the continent. American readers will understand the allusion to Lewis and Clark; this note is added to help every reader to grasp the implications of the development of North America after being a largely untrodden land less than 150 years ago.

† The book by Ayres and Scarlott, mentioned on page 163, gives an interesting account of projects for warming houses by stored solar heat.

‡ 'The people asked for bread, and were offered incombustibles. *Cf. Matthew* vii, 9.

MISCELLANY SIX

Among the Trees

THOUGH leguminous trees are common in tropical and sub-tropical forests and scrubs (acacias were plentiful at Kongwa, and the 'wattle' – a legume – is Australia's national emblem) the presence of legumes is not indispensable for growth or cropping. An absence of legumes, whether that absence is a natural one or is due to conditions unsuitable for growth, does limit the kind of cropping which can be undertaken. Thus, British forestry makes no use of legumes, since our largest leguminous plant is gorse, whin, or furze. There is no leguminous tree larger than the laburnum, or the Judas tree, able to grow well in Britain. There was no native leguminous pasture plant in Australia or New Zealand. The poverty of native large animals in those countries may not be unconnected with that.

Besides forestry, there is tree-agriculture. The difference between these is that forestry grows trees primarily for timber, whereas tree-agriculture grows trees for products which can be obtained repeatedly during the life of each tree. Rubber, cacao, coconuts, and palm oil are examples of crops from tree-agriculture.

Tree-agriculture is often independent of leguminous plants. This is well shown in Apulia, where hundreds of square miles of soil lying a few inches deep on a limestone pavement are covered with olives and almond trees and occasional vineyards. The soil is too thin to be ploughed, and has to be cultivated with mattocks. It is also too little retentive of moisture for any annual crop to be grown, except a catch crop of beans for human food, taken in odd corners in spring before the leaves of the trees are fully developed. There are consequently no fields in our sense. The region provides oil and beans for food, and oil and almonds for sale to buy wheat and other necessaries.

It is worth noting that though the soil of Apulia is not devoid of lime, the most popular and most effective nitrogenous fertilizer in that region is calcium cyanamide, a lime-containing synthetic fertilizer which is not made in Britain at all, because its manufacture requires ample electric power. It is not solely on domestic economic grounds that the apparently paradoxical choice of cyanamide is made for this lime-rich region; it is mainly for reasons connected with actions of soil bacteria upon the insoluble cyanamide when that is moistened by rain, causing ammonia to be slowly released from the fertilizer. Ordinary nitrogenous fertilizers would be likely to be washed out of the soil by a single fall of rain. Here – as in many semi-arid areas – rainfall, though infrequent, is apt to be heavy when it does come.

This example may suggest that planners who look no further than more-nitrogen-for-more-production should beware of complexities which may call for a good deal more ingenuity than appears on the face of things.

The soil microbes and the climate *cannot* be left out of account.

There is another form of agriculture associated with trees. This is shifting cultivation. It is practised in the central parts of Africa and in parts of the Orient, and in other regions covered with scrub and jungle. The dense vegetation of such places has often misled Westerners into admiration of the great 'fertility' of the land. The rapid growth of trees and the high productiveness of wood (one can hardly say of *timber*) so long as the ground is covered with forest or with the ash of the trees, reflects the condition that the greater part of the plant-nutrient capital is *above* the soil: not in the soil, but locked up in the forest growth itself, and in the forest litter. There is a high rate of turnover of a limited amount of capital, and most of that turnover goes on above the ground, when trees and lianas (as well as leaves) fall bodily to be worked over by the insects and other small fauna and finally by the true microbes. With the prevailing high rainfall and temperatures, and under almost perpetual shade, these operations of decomposition of effete material and restoration of the nutrients to circulation occur very rapidly.

Drastic changes are made by removing the trees as timber. Apart from any effect on the people, two things happen: the plant nutrients in the trees are carried away, and the ground is bared to the destructive influences of intense sun and rain.

Shifting cultivation is commonly practised on soils derived from rocks poor in nutrients; often, these rocks are very old and tough, weathering only slowly to release what nutrients they do contain. Attempts on Western lines to get cleared land to bear annual crops are likely to fail after a season or two.

Shifting cultivation consists essentially of felling and burning the jungle *in situ*, snatching two or three poor crops, and then allowing the site to revert to jungle. Reversion to forest occurs within fifteen or twenty years under those conditions; even so, the rotation is a long one, and occupies a good deal of land. It is essentially a rotation of wild trees (twenty years) and crops (three years). This rotation may not be able to support a large population on a given area, but a better and more productive system of agriculture for those soils has not been devised.

Shifting cultivation has been roundly condemned by people who have summarily and unthinkingly transported to unfamiliar circumstances a mentality gained from European forestry and agriculture performed in totally different situations. The shifting-cultivation forests are not well adapted to the winning of timber (even if one overlooks the circumstance that to clear the forests, or to reserve them for timber, might deprive the natives of a living); they are still less well suited for cropping, European-style.

Objections about the 'wastefulness' of shifting cultivation are misconceived. The best and most rational, and also the most humane, use of these jungles is precisely that which the natives have evolved: namely, to adopt a long rotation of crops and trees.

It may be that liberal lawyers and other well-meaning

people will one day begin to look for realistic remedies for real causes of discontent, and will no longer be satisfied to advocate giving black and brown men a vote and/or to believe that everything will be lovely from the time when an impeccably-drafted Charter of Human Rights has at last been drawn up. There will still be the land and the crops and the trees; and a seemingly increasing population contesting for all those and not to be stayed by a Formula or even a Form.

SUMMARY

RECENTLY, the physical sciences have tended to be unduly prominent in the popular picture of what science and technology can mean for living.

We must eat in order to live.

To stress the need for diffusing knowledge about the limitations of chemistry and soil microbiology is not treachery to those central sciences of living. It means just that chemistry, and all large organisms such as man, can operate only so long and so far as there exists *substance* capable of being applied to reduce atmospheric oxygen. External energy is not enough. Unless translatable and translated into terms of food chemistry, the possibilities of energy are only potential and secondary for purposes of living.

This suggests an increased importance for the sciences collectively called 'agricultural', since we live by means of oxidation and reduction, and not simply in terms of energy.

Beyond the need for something to reduce oxygen gas, there is an additional stringent requirement, or set of requirements, to be satisfied whenever chemistry or physics is sought to be brought into relation with food-production. It is: that there must also be reduced nitrogen in suitable form and ratio.

Photosynthesis is concerned with build-up of carbon and water only. It does not, in general, 'take care of' nitrogen gas.

Food-production is largely a matter of transformation of the three atmospheric gases, carbon dioxide, oxygen, and nitrogen. The only generally satisfactory method of bringing nitrogen in suitable kind and adequate proportion into foodstuffs from the land consists of cultivation of leguminous plants. This invokes the legume nodule bacteria; it implies that they, and other soil microbes, must be sustained at an appropriate level of activity: through use of earthy substances additional to those which exist under natural conditions of equilibrium of the thin smear of top-soil.

Summary

The extent of appeal to calcium, sulphate, potassium, phosphate, and other earthy matter may be relatively small and, if maintenance of a small human population will suffice, it need not extend to fossil fuel. Nevertheless, earthy materials are indispensable for any mode of manipulation of atmospheric gases into food.

A very ponderable call upon earthy substances is a *sine qua non* of greater production of food.

This strain upon mineral resources (including ores) is likely to be appreciably increased by tank-farming techniques. However, all methods for increasing the effective output of food above the present level imply use of earthy substances as plant nutrients and soil amendments, *plus* whatever is applied to make metal or 'synthetic' implements and apparatus, and to provide power for using them.

The alternative is to adopt the traditional Chinese mode of subsistence living, subject to natural checks. The alternative will not necessarily be voluntary: it will impose itself, as fossil fuels and/or metallic ores become exhausted.

That is the position in which mankind finds itself for the first time, after a historically brief adventure in utilization of fossil resources. This escapade has been motivated partly by a desire for amenities (such as rapid communication) in parallel with an unprecedented increase in output of food during the last half-century.

The several uses of fuel are competitive; and need regulation.

No organization is concerning itself with the food-and-fuel problem as a whole. Mankind being under the delusion that if it has not yet conquered nature, great progress has been made that way, man is in reality letting things take a natural course. The effect has been to accelerate a process. The natural fluctuations of population have been exaggerated for a while: that is all.

APOLOGUE

FEW people of any pretensions read Jules Verne nowadays. It is a little old-fashioned to think of crossing Africa in a balloon, and the hazard of falling among cannibals no longer stirs the blood of adventurous stay-at-homes. Verne's geology may be questionable at times, as when he finds coal in the middle of an extinct volcano; and electric light and electrical marine propulsion are no longer the novelties they were when the submarine *Nautilus* was doing its more than twenty thousand leagues in 1866. Still, if you grant Jules Verne his premisses, there is nothing much wrong with the results.

Captain Nemo (we are told) explained: 'Recollect this. I owe everything to the ocean. It produces electricity, and electricity gives the *Nautilus* heat, light, speed, and, in a word, life!'

It is not clear that Captain Nemo cultivated anything, except pearls. The ocean contained more than enough for the needs of some involuntary passengers and a crew, who altogether cannot have numbered more than a hundred. Captain Nemo ranged wherever he wished, and laid under tribute the whole resources of the sea.

Electrical energy was 'the very soul of his mechanical arrangements'. It gave motion to the ship, and kept its crew alive. But, suppose the *Nautilus* had been immobilized through lack of fuel . . . ? Could its inhabitants have been kept alive on what they could gather, or cultivate, on a spot remote from the mine of pit-coal, or sea-coal (the translator may have felt a wee difficulty, there) which was used to extract sodium from the sea-water and thus to generate electricity chemically from batteries?

The plants and animals which gave the people of the *Nautilus* food, clothing, and cigars were, it is true, wild and uncultivated; but they would mostly have been beyond reach if the ship had not been able to revive her batteries through the medium of coal. It was fuel – in our plain, everyday, ordinary sense – which enabled the microcosm of the *Nautilus* to feed and live.

Different as the method of the *Nautilus* was from our customary mode of living, the principle of dependence upon fossil fuel for a decent, tranquil, and integrated life is the same. Was it, I wonder, by accident or intention that Jules Verne hit upon that parable? Or this one:

In the *Nautilus*, every knife, fork, spoon, and plate bore a motto surrounding a letter:

That is one apologue, and it is tempting to stop there, leaving the chemical symbol for nitrogen flaunting a final hint. That finish might be obscure; but, even if interpreted rightly, it would do no more than indicate the material of a possible and partial remedy. It could not convey a suggestion about the underlying state and perdurable condition which is the same for microbes as for men.

For another apologue let us go back to the Heroic age of classical Greece, and borrow from the umbrageous time when thought about the condition of man was emerging from myth into a beginning at cosmology. I cite Heraclitus in the summary given by Professor Benjamin Farrington (*Greek Science*, Penguin Books, 1944). Heraclitus 'was the philosopher of change. His doctrine has been summed up in the phrase *Everything flows*; but his choice of Fire as his First Principle was probably not due, as is often said, to its being the most impermanent of things, but to its being the active agent which produces change in so many technical and natural processes. Still more important was his idea of Tension, brought in to explain the relative permanence and fundamental impermanence of things. It is one of the richest and most helpful ideas of the old philosophers, not a whit reduced in significance when we remember that it, too, had its origin in the techniques of the time. . . . According to Heraclitus there is in things a force that moves them on the Upward Path to Fire, and an opposite force that moves them on the Downward Path to Earth. The existence of matter in any particular state is the result of a balance of opposing forces of Tension. Even the most stable things in appearance are the battleground of opposing forces, and their stability is only relative. Nature as a whole is either on the Upward Path to Fire or on the Downward Path to Earth. Its mode of existence is an eternal oscillation between these two extremes.'

Accepting Professor Farrington's caution about the danger of reading into the ideas of old thinkers the meanings of a later age, I dismiss the reference to the prominent positions of Fire and Earth as no more than interesting, however germane. What I wish to point out as significant is the reference to what Professor Farrington called 'tension'.

I venture to suggest that in this idea Heraclitus may have sought to impart the most fundamental of principles in natural history: namely, the notion of equilibration. The sense is unchanged – I think it is reinforced – if for Tension we read Equilibrium. This idea of equilibration of forces may well have come from consideration of the string of a bow or a lyre. The string and the frame which supports it must be subject to equal and opposing forces, else one or both will break, or be slack. *Ut tensio, sic vis.*

Heraclitus can have had no inkling of biological balance; it was natural that the notion of equilibration of dynamic strains should have come to fruition through a flexible object such as a bow. Only in our own time is the notion of equilibration sufficiently familiar for it to be adaptable to human communities.

It is not yet common, since man often forgets that he also is a part of nature. Sometimes he appears to act as if he thought he had put himself above and outside the operations of nature. Sometimes the principle of equilibration is, as it were, rusticated; and accepted only with difficulty when brought to light. Medical bacteriologists, for example, seem generally to prefer a study of individual species of 'attacking' microbes to a view which would be sounder and more ecological by tending more to take into consideration not only the susceptibilities of the attacked but also the conditions governing the resultant of the behaviour of the attack-defence complex. The great breadth of view expressed by Dr René Dubos (page 156) lends colour to the thought that what many medical and veterinary people mistake for the fundamentals of microbiology are specialized aspects of it. Pasteur, venerated by microbiologists and chemists everywhere, has had some of his pertinent truths about microbes distorted, and others shelved, by the majority of those who pretend to be his disciples.

Some people, some animals, and some plants will die or suffer because of the effects of microbes; but all plants, all animals, and all people *live* because of the effects of soil microbes.

It follows – without necessary proof – that men can live in numbers on the earth only in proportion to what they can suscitate from soil microbes. Mankind is not apart from ecology: but is part of a soil-based microbiological association. So long as man has used the fossil products of the earth in an unreflective and hollow attempt to keep the natural equilibria at bay, so has he increasingly felt himself to be clever, civilized, 'scientific'; and so long have there been disquisitions with titles like 'The Romance of Petroleum', 'Man's Conquest of the Earth' (or 'of Nature'), 'Great Engineering Feats', and 'Chemistry the Servant of Man'. So far, so good; but good only so far as the various kinds of technological achievement can be kept up.

Perhaps they can be kept up beyond the limits which at present seem likely: not necessarily in their present forms, but serviceably in ways we cannot foresee. If so, it will be good; but always and inevitably subject to the condition of maintaining a state of equilibrium. Knowing about equilibrium does not mean recourse to planning in the present-day, almost exclusively deterministic, sense: for that is unscientific and is practised by and under the direction of non-scientists. When man takes full account of equilibria and of the roles played by chance in determining them, he can call himself scientific. It is important not to confuse technology with science in the broadest and most powerful connotation.

If the future equilibrium between man and his world is not natural as it approximately was a century or two ago, the equilibrium may be artificial, as it now is; but equilibrium between man and soil microbe there must be, since that is a prime condition of existence. There is no dodging that conclusion. The equilibrium or Tension may be altered: its existence cannot be ignored, unless the string or bow breaks or becomes slack and useless.

It is timely to begin to recognize that this vital Tension exists, and to enquire about what it consists of and how it is maintained. Have we been too little curious about it, too long?

* * *

FOR an ending I cannot do better than quote the question of a Scottish married lady without scientific training or other apparent special aptitude outside, presumably, housewifery. We were returning to Glasgow across the Clyde from a Workers' Educational Association week-end School at Dunoon, where I had been the lecturer and had outlined the relations between food, fuel, and microbes. It was pure coincidence that our vessel was the electrically-driven *Talisman*.

This member of the School found me on deck, and said, hesitantly: 'In view of what you were saying about the microbes and their struggles and ups and downs in population and so on, depending on what they find in the soil, would it be right to think that the same sort of thing applies to people – that they can get out of adjustment at times, but only for a little while perhaps, until things right themselves somehow – perhaps by deaths if no more food is to be found?'

I replied: 'Yes, I think that is a very good summary of the position. People are naturally not above natural operations any more than the soil microbes are. Only, man has lately learnt to support himself in numbers far above the natural level of population. He can go on doing this by applying intelligence; that is, by putting himself in a quite artificial relationship to his surroundings. He makes a bet with himself that he can keep that up.

'So long as he uses intelligence, there may be no need to worry about the outcome. But if, for any reason, people fail to keep this artificial balance, they will be subject to the same ecological principles as the microbes. There can be no escape except by continuously applying intelligence.

'It's something like swimming. The operations of density and solution are the same for all organisms. If you learn to swim and to keep your head above water, you are all right. But if for any reason your head is not kept above water – whether because you have lost your lifebelt or for any other reason – the result must be the same.'

Each person learns for himself or herself. One person can do little, but, within certain limits, man collectively can control destiny.

* * *

ACKNOWLEDGING throughout that any mistakes, or misinterpretations of information or advice tendered, are solely mine, it is with pleasure that I thank the colleagues and friends who have criticized parts of this book in draft or have contributed to it. Especially are thanks due to Professor W. W. Fletcher, of the Botany Department of the West of Scotland Agricultural College, for his help with the botany. I am also indebted to Professor Matthew Meiklejohn, of the Department of Italian at Glasgow University, for supplying a gloss of Carducci's latinisms. Professor Meiklejohn's sister is a well-known soil bacteriologist.

BOTANICAL NAMES OF SOME PLANTS MENTIONED

LEGUMES

acacias: including wattle – *Acacia* spp.
beans, various – *Vicia faba*, *Phaseolus*, *Dolichos*, and others
clover, Alexandrian (berseem) – *Trifolium alexandrinum*
  'annual' red (cow-grass) – *T. pratense*
  'annual' white (Dutch white) – *T. repens* var.
  crimson – *T. incarnatum*
  perennial ('wild white') – *T. repens*
  subterranean[1] – *T. subterraneum*
eggs-and-bacon, bird's foot trefoil – *Lotus corniculatus*
ground nut (peanut) – *Arachis hypogea*
indigo – *Indigofera tinctoria*, etc.
kudzu – *Thunbergia pueraria*
lespedeza – *Lespedeza sericata*
lucerne (alfalfa) – *Medicago sativa*
peas, edible, various – *Pisum sativum*, *Cicer*, and others including *Lathyrus*
peas, sweet – *Lathyrus* sp.
shamrock[2] – *Trifolium repens*
soya bean, soybean – *Soja max* (*Glycine max*)

1. Not specifically mentioned in text; is the principal clover used in Australia.
2. Originally *Oxalis acetosella* (sorrel) – a non-legume.

Clovers and clover-like crop plants are numerous, including the medicks. Sweet clover (*Melilotus*) is not a true clover but resembles lucerne.

NON-LEGUMES

algae, of soil; also seaweeds
almond – *Prunus amygdalus*
cabbage tribe: broccoli, cabbage, cauliflower, kale – *Brassica oleraceae*
cacao (cocoa) – *Theobroma cacao*
cereals:
  barley – *Hordeum sativum*
  maize: including hybrids – *Zea mays*
  oats – *Avena sativa*
  rice: *padi* or 'wet' – *Oryza sativa*
  wheat, hard and soft – *Triticum vulgare*
cotton – *Gossypium* spp.
grasses, true: including cereals and sugar-cane, *q.v.*
  smooth-stalked meadow grass (blue grass) – *Poa pratense*
  timothy – *Phleum pratense*
hop – *Humulus lupulus*
mangold (mangel wurzel) – *Beta vulgaris*
mulberry – *Morus alba*
olive – *Olea europaea*
potato – *Solanum tuberosum*
rubber – *Hevea brasiliensis*
sugar-beet – *Beta maritima* or *B. rapa*
sugar-cane – *Saccharum officinarum*
tea – *Thea sinensis*
turnip: including swede (rutabaga) – *Brassica rapa*
vine (grape-) – *Vitis vinifera*
yeast, baker's or 'food' – *Saccharomyces* spp. and others

INDEX

Subjects and places mentioned only allusively, *e.g.* Ardrossan (p. 229), New York State (p. 189), are not all entered below.

Index

Index